Agricultural Marketing

NIPA® GENX ELECTRONIC RESOURCES & SOLUTIONS P. LTD.
New Delhi-110 034

Agricultural Marketing

MCQ's for JRF, SRF, ARS, NET, SET
Civil Services and Other Competitive Examinations

B. Kavitha
Post-Doctoral Fellow
Department of Agricultural Economics
Tamil Nadu Agricultural University
Coimbatore, Tamil Nadu, India

M. Uma Gowri
Post-Doctoral Fellow
Department of Agricultural Economics
Tamil Nadu Agricultural University
Coimbatore, Tamil Nadu, India

A. Indhushree
Senior Research Fellow
Department of Agricultural Economics
Tamil Nadu Agricultural University
Coimbatore, Tamil Nadu, India

S. Padma Rani
Associate Professor
Department of Agricultural Economics
Tamil Nadu Agricultural University
Coimbatore, Tamil Nadu, India

NIPA® GENX ELECTRONIC RESOURCES & SOLUTIONS P. LTD.
New Delhi-110 034

NIPA® GENX ELECTRONIC
RESOURCES & SOLUTIONS P. LTD.

101,103, Vikas Surya Plaza, CU Block
L.S.C. Market, Pitam Pura, New Delhi-110 034
Ph : +91 11 27341616, 27341717, 27341718
E-mail: newindiapublishingagency@gmail.com
www: www.nipabooks.com

For customer assistance, please contact
Phone: + 91-11-27 34 17 17
Fax: + 91-11-27 34 16 16
E-Mail: feedbacks@nipabooks.com

ISBN: 978-93-95319-59-1

Composed and Designed by NIPA®.

Preface

This book "Agricultural Marketing" is envisioned to be the primary objective book of this subject with a vision to helps in preparation especially for competitive examination like Junior Research Fellowship (JRF), Senior Research Fellowship (SRF), Agricultural Research Service (ARS), National Eligibility Test (NET), State Eligibility Test (SET) and Civil Services too. The book is structured fundamentally with objective type questions at a student level covering all the areas of Agricultural Marketing. Chapter wise questions begin with student learning objectives to meet the broad learning outcomes. Moreover, students are encouraged to pay attention to these objectives type questions as they go through the content of the subject. At the same time, this book includes a number of current topics designed to enhance rapid learning process of the students. Many students treat the book as a handy guide to assess their knowledge in the field of Agricultural Marketing across the country.

Authors

Contents

1

Market: Components of Market Importance of Agricultural Marketing

1. **The word 'market' is a derivative of the Latin word ————**

   a) Marcatus b) Markus

   c) Marcat d) Martus

2. **'Marcatus' is a ———— word.**

   a) Greek **b) Latin**

   c) French d) Spanish

3. **The word 'Marcatus' means ————**

   a) Merchandise b) Merchant

   c) Trader d) Money

4. **———— implies a chain of activities taken in moving the goods from the production point to the consumption point.**

   a) Supply b) Sale

   c) Marketing d) Stocking

5. **From the marketing system, ———— want the maximum possible price for the produce produced at farm.**

   a) Consumer **b) Farmers**

   c) Government d) All the above

6. **____________ comprises all the information, facts, data, opinions and figures which influence the marketing of goods and services**

   a) Market information b) Market news

   c) Market arrivals d) None of the above

7. **____________ comprises of information relating to the price that prevailed in the past period and market arrivals over time.**

   a) Market intelligence b) Market news

   c) Market information d) Marketing

8. Market conduct includes

a) Policy towards setting price

b) Policy towards setting product quality

c) Policy towards coercing rivals

d) Policy towards adjusting size and number of firms

9. Key elements of market conduct are

a) Firm objectives,

b) Dynamic progressiveness of the system

c) collusive behaviour,

d) Competition methods

10. The patterns of firm's behaviour of especially in relation to pricing and their practices

a) Market performance **b) Market conduct**

c) Market structure d) Market efficiency

11. The effectiveness of sellers/suppliers in a market by utilizing the economic resources of the market for the ultimate benefit of consumers/ buyers is called ____________________

a) Market performance b) Market conduct

c) Marketing efficiency d) Market structure

12. Composition of end results in the dimensions of price, output, production cost, selling cost, product design is

a) Market performance b) Market integration

c) Market conduct d) Market structure

13. ____________ does not come under the components of market conduct

a) Dynamic progressiveness of the system

b) Competition methods

c) Product differentiation

d) Firm objectives

14. The organizational features of a market are called ______________.

a) Market performance **b) Market structure**

c) Market conduct d) Marketing Channel

15. ____________ is not a type of market structure?

a) Competitive monopoly b) Oligopoly

c) Perfect competition d) Functioneries.

16. If the market demand curve for a commodity has a negative slope then the market structure must be

a) Perfect competition

b) Monopoly

c) The market structure cannot be determined from the information given.

d) Imperfect competition

18. __________ is a characteristic of monopolistic competition?

a) A differentiated product

b) Easy entry into and exit from the industry.

c) Few sellers

d) All of the above are characteristics of monopolistic competition.

19. In which form of market structure would price be the key factor when competing?

a) Monopoly b) Oligopoly

c) Monopolistic competition **d) Perfect competition**

20. Product variation refers to

a) An activity undertaken by a firm to make demand more price inelastic.

b) A problem with quality control that tends to decrease demand.

c) An activity undertaken by a firm to increase demand.

d) None of these.

21. The economic results that flow from the industry that each firm in that industry pursues its particular line of conduct.

a) Market efficiency b) Market conduct

c) Market structure **d) Market performance**

22. Criteria for measuring market performance and efficiency of the market structure is________

a) Resource use efficiency including real cost

b) Monopoly profits including the average cost

c) Inventing new products to maximize individual welfare.

d) Whether or not the system aggravates the problem of inequalities.

23. The firm's conduct depends on __________ aspects of market structure

a) The number of buyers and sellers

b) The existence and degree of barriers to entry

c) Economies of scale

d) The degree of product homogeneity

24. The behavior or comportment of buyers and sellers to the market structure is ________

a) Market performance
b) Market efficiency

c) Market conduct
d) Market structure

25. In ________ duopoly, each firm assumes that other will not change the price in response to its price changes.

a) Bertrand
b) Ricardian

c) Strackelberg
d) Cournot

26. _________ market has more number of sellers and small number of buyers

a) Oligopoly
b) Oligopsony

c) Duopoly
d) Monopsony

27. ________ type of duopoly where both two firms assume each others output and consider it as a fixed quantity and produce it in their own firm.

a) Cournot
b) Ricardian

c) Bertrand
d) Strackelberg

28. Market conduct does not comprise of

a) Policies towards market sharing and price setting

b) **Policies towards promoting co-operative tendency**

c) Policies aimed at coercing rivals

d) Policies towards setting the quality of products

29. Perfect competition is not characterized by

a) Free entry into or exit from the industry

b) Knowledgeable buyers and sellers with respect to prices.

c) Large number of buyers and sellers

d) **Considerable advertising by individual firms**

30. Which market types has a large number of firms that sell similar products but slightly different?

a) Oligopsony
b) Perfect competition
c) Oligopoly
d) Monopolistic competition

31. The product homogeneity resulting from grading can bring the market as ____________

a) Monopoly
b) Perfect competition
c) Monopolistic competition
d) Monopsony

32. ________is not one of the market components

a) A product for transaction
b) Demarcation of area
c) Buyers and sellers
d) Time

33. High degree of market concentration generates____________ situation in the market.

a) Monopolistic
b) Monopoly or Monopsony
c) oligopoly or oligopsony
d) Monopsony

34. The patterns of behaviour of firms indicates

a) Market Structure
b) Market Conduct
c) Market Performance
d) Market Behaviour

35. Market conduct includes

a) Policies towards Market sharing and price setting
b) Policies aimed at coercing rivals and
c) Policies towards setting the quality of products
d) All of the above

36. The market structure determines

a) Market conduct
b) Market Performance
c) Both
d) None of the above

37. Behaviour of market with regard to price determination, promotion of sales and government regulation is called

a) Market Structure
b) Market Conduct
c) Market Performance
d) Market Behaviour

38. The economic result of market structure and market conduct is called

a) Market conduct **b) Market Performance**

c) Market Structure d) None of the above

39. Market performance resembles

a) Level of Price & Investment b) Profit Margin

c) Reinvestment of Profit **d) All the above**

40. Market conduct includes

a) Market sharing and price setting policies

b) Policies aimed at coercing rivals

c) Policies setting the quality standards

d) Efficiency in the use of resources

41. Market information includes

a) Market intelligence b) Market news

c) Both a and b d) Marketing

42. Historical nature of market information is

a) Market intelligence b) Market news

c) Market information d) Marketing

43. Current information about markets is

a) Market intelligence **b) Market news**

c) Market information d) Marketing

44. ____________ is defined as gathering, recording and analyzing all the facts related to the transfer and sale of goods and services from producer to consumer

a) Market intelligence b) Market news

c) Market information **d) Market research**

45. It is not the step of market research

a) Hypothesis formulation

b) Problem identification

c) Data analysis

d) Undesigning the empirical procedures

46. _____________ may be broadly defined as a communication or reception of knowledge or intelligence.

a) Market information b) Market news

c) Market arrivals d) None of the above

47. _____________ describes research on markets, market size, geographical distribution and incomes

a) Market research b) Marker conduct

c) Market information d) Market news

48. Major objectives of market research

a) Understanding the existing traditional marketing system

b) Diagnosing the problems in a dynamic content

c) Analyzing and predict the impact and effectiveness of alternate policy measures

d) All the above

49. Which of the following is/are the components of market structure?

a) Concentration of market power

b) Degree of integration

c) Degree of product differentiation

d) All the above

50. The market information pertaining to market arrivals, prices, demand for commodities etc., in the past period refers to

a) Market news **b) Market intelligence**

c) Communication d) Market extension

51. The market information pertaining to market arrivals, prices, demand for commodities etc., in the current period refers to

a) Market news b) Market intelligence

c) Communication d) Market extension

52. _____________ can be measured by a number of factors, such as the number of competitors in an industry, the heterogeneity of product and the cost of entry and exit

a) Market conduct b) Market intelligence

c) Communication d) Market news

53. __________ is applied to test the viability of a new product or service by communicating directly with a potential customer

a) Market information

b) Marker intelligence

c) Market research

d) Market news

54. The concentration of market power is an essential element in determining the nature of competition & consequently of _______ and __________

a) Market information and market news

b) Marker intelligence and market information

c) Market conduct and performance

d) Market news and market intelligence

55. _________ does not come under the components of the market structure and does not determine the conduct and performance of the markets are

a) Concentration of market power

b) Degree of product differentiation

c) Conditions for entry of firms in the market

d) Technology

56. Marketing research is concerned with _________

a) Anticipation of production

b) Supply position.

c) Financial problems.

d) Solution to specific problems of marketing

57. The process of subdividing total markets into several sub market is __________

a) Market fluctuations

b) Market positioning

c) Market segmentation

d) Market penetration

58. Mercatus means _________

a) Buying

b) To sell

c) To assemble

d) To trade

59. Markets are created by________

a) Nature **b) Economic force**

c) Businessmen d) Product

60. Fixing a high price for a new product will be called as ________

a) Price skimming b) Price segmentation

c) Dual pricing d) Customary pricing

61. _________ refers to the size and design of the mar-ket including the manner of the operation of the market.

a) Market structure b) Market conduct

c) Both a & b d) None of the above

62. _________ is a component of market structure.

a) Concentration of market power

b) Flow of market information

c) Degree of integration

d) All the above

63. A ________degree of market concentration restricts the movement of goods between buyers and sellers at fair and competitive prices.

a) High b) Low

c) Not affected d) None of the above

64. The price variations in the market will not be wide-ranging for — _______ products.

a) Homogeneous b) Heterogeneous

c) Both a & b d) None of the above

65. When products in the market are __________, firms tended to set different prices for their products.

a) Homogeneous **b) Heterogeneous**

c) Both a & b d) None of the above

66. __________ is a component of market structure.

a) Degree of Product Differentiation

b) Conditions for entry of firms in the Market

c) Degree of Integration

d) All the above

67. _________ is a significant element in determining the nature of competition, market conduct and performance.

a) Concentration of market power

b) Flow of market information

c) Degree of integration

d) All the above

68. __________ refers to the market characteristics which affect the traders' behaviour and their performances.

a) Market structure b) Market conduct

c) Both a & b d) None of the above

69. The prime of object of production is its ________

a) Consumption b) Distribution

c) Exchange d) None

70. ________ is the advantage of selling through direct channels.

a) Simple and fast

b) Economical

c) Reduce dependence on middle men

d) All the above

71. Channel in which product is sold through wholesalers, retailers and agents is called ________

a) Direct channel **b) Indirect channel**

c) Both a & b d) None of the above

72. The behaviour patterns of firms indicates

a) Market Structure **b) Market Conduct**

c) Market Performance d) Market Behaviour

73. The market structure determines

a) Market conduct b) Market Performance

c) Both d) None of the above

74. Behaviour of market with regard to price determination, promotion of sales and government regulation is called

a) Market Structure **b) Market Conduct**

c) Market Performance d) Market Behaviour

75. The economic result of market structure and market conduct is called

a) Market conduct **b) Market Performance**

c) Market Structure d) None of the above

76. Market performance resembles

a) Level of Price & Investment b) Profit Margin

c) Reinvestment of Profit **d) All the above**

77. Creation of utility by way of marketing is known as

a) Distribution b) Production

c) Exchange d) Consumption

78. The exchange of goods and services expressed in terms of money is referred as ________

a) Marketing **b) Price**

c) Transaction d) Product

79. __________ is creating the quality specifications of the uniform grades of products among buyers and sellers over space and over time

a) Standardization b) Packaging

c) Processing d) Storage

80. Composition of end results in the output, product design, dimensions of price, selling cost and production and is called

a) Market integration b) Market conduct

c) Market structure **d) Market performance**

81. Market information means________

a) Knowledge of industries

b) Knowledge of households

c) Knowledge of peers

d) Knowledge of customer's tastes

82. Marketing is the art of__________

a) Buying more b) Paying more

c) Selling more d) Talking more

83. Marketing is a_________

a) One day effort **b) Team effort**

c) One man effort d) All the above

84. Factors influences marketing __________

a) Product demand
b) Public taste
c) Buyer behaviour
d) All of the above

85. Marketing helps in __________

a) Boosting production
b) Getting new clients
c) Interacting with strangers
d) All of these

86. Selling is _________

a) Different from marketing
b) A sub-function of marketing
c) Same as marketing
d) None of these

87. The long term objective of marketing is_________

a) Customer Satisfaction
b) Profit Maximisation
c) Cost cutting
d) Profit Maximisation with customer

88. Subject matter of Marketing?

a) Machine, man and money
b) Product or Service
c) Capital
d) Brand

89. Marketing is_________

a) A day to day function
b) A one-off affair
c) A one man show
d) A collective process

90. Which is the subject Matter of marketing?

a) Man
b) Labour
c) Planning
d) Goods and Services

91. Marketing plan helps in_________

a) Better lead generation
b) Better systems
c) Better results
d) Improved balance sheet

92. When a product gets an attractive name, symbol or any identity mark to look different from other products is known as

a) Packaging
b) Branding
c) Pricing
d) Marketing
e) Assembling

2

Classification of Markets and Marketing Functions

1. **Village market is based on the classification of ______________**

   a) Volume of transaction **b) Location**

   c) Time span d) Area coverage

2. **Regional Markets are based on __________**

   a) Location b) Time span

   c) Area coverage d) Volume of transaction

3. **The prices of commodities are controlled mainly by the extent of demand for the commodity rather than the supply of the commodity.**

   a) Secular Markets

   b) Forward Markets Secular Markets

   c) Long-period Markets

   d) Short-period Markets

4. **__________ is a type of markets where the major transaction of commodities takes place between the village traders and wholesalers.**

   a) Village Markets

   b) Primary Wholesale Markets

   c) Secondary Wholesale Markets

   d) Regional Markets

5. **Primary wholesale market and secondary wholesale market are under the category of**

   a) Area coverage b) Volume of transaction

   c) Time span **d) None**

6. **Markets for manufactured goods are known as________**

   a) **Secular Markets** b) Wholesale Markets

   c) General Markets d) Long-period Markets

7. **In____________, goods are exchanged for cash immediately after the sale is completed**

   a) Forward Markets **b) Spot or Cash Markets**

   c) Perfect Markets d) Capital Markets

8. **If purchase and sale of a commodity takes place at time 't' but the exchange of the commodity takes place in future time t + 1, it is referred as ___________**

   a) Long-period Markets b) Capital Markets

   c) Retail Markets **d) Forward Markets**

9. **Markets for car is**

   a) World Market b) National Markets

   c) Regional Markets d) Short-period Markets

10. **Markets based on time span is**

    a) Secular Markets b) Spot markets

    c) Forward markets d) Terminal markets

11. **Market for flower is _____________**

    a) Specialized Markets b) Secular market

    c) Rural Market d) Village market

12. **_______ is the process of obtaining a required product from someone by giving something in return.**

    a) Supply **b) Exchange**

    c) Selling d) Marketing

13. **Dissemination of product details viz, price, quantity, locality, etc through multi media is known as__________**

    a) Physical Movement Functions **b) Facilitative Functions**

    c) Exchange Functions d) Management Functions

14. **Many activities performed in carrying a good or commodity from its production point to its ultimate consumption point is termed as ________________**

    a) Marketing function b) Market channel

    c) Market route d) Market functionaries

15. Which is the most prior function performed in the marketing of agricultural commodities

a) Packaging b) Storing

c) Harvesting d) Distribution

16. Creation of utility by way of marketing is known as

a) Distribution b) Production

c) Exchange d) Consumption

17. Decisions are continuously made in the day to day operations of farm business and involve small investment are known as

a) Strategic decisions b) Marketing decisions

c) Operational decisions d) None of the above

18. Decisions regarding irrigation, conservation and reclamation programmes are comes under

a) Strategic decisions b) Marketing decisions

c) Operational decisions d) None of the above

19. Decisions regarding selection of enterprise are known as

a) Strategic decisions b) Marketing decisions

c) Operational decisions **d) Administrative decisions**

20. Management decisions, which involve lasting effects, are known as

a) Strategic planning b) Marketing decisions

c) Operational decisions d) Administrative decisions

21. Standardization of Grades are ________ marketing function

a) Facilitative b) Exchange

c) Physical d) All of the above

22. Storage and warehousing are _________ marketing function

a) Facilitative b) Exchange

c) Physical d) All of the above

23. Buying and selling are ________ marketing function

a) Facilitative **b) Exchange**

c) Physical d) All of the above

24. The facilitating functions of marketing are as i) transportation ii) financing iii) Dissemination of Market Information iv) insurance

a) i, ii and iii are correct b) i, ii and I are correct

c) i, iii and iv are correct **d) ii, iii and iv are correct**

25. Which statement is correct?

i) Standardization follows grading

ii) Grading follows standardization

iii) Grading is a sub-function of standardization

iv) Standardization is a sub-function of grading

a) i and ii b) i and iv

c) ii and iii d) ii and iv

26. _______________ is a kind of marketing function by which placing goods into attractive packets and containers according to the customer continence

a) Packaging b) Branding

c) Pricing d) Marketing

27. Function of marketing associated with the development of goods with respect to their color, design, shape and other characteristics is called______________________

a) Production

b) Product Planning & Development

c) Packaging and Pricing

d) Standardization and Grading

30. The Organization which deals with the purchasing of goods for resale. These manufacturing Organizations buy raw materials and components. This marketing function is called as ________

a) Product Planning and Development

b) Packaging and Pricing

c) Standardization and Grading

d) Buying and Assembling

31. The Marketing Function deals with cost of the product and the capacity of consumer to pay is called

a) Packaging b) Branding

c) Pricing d) Marketing

3

Market Functionaries Regulated Market and Cooperative Market

1. **____________ aims at elimination of unhealthy and unscrupulous practices, reducing the marketing charges and providing facilities to producer and sellers in the market**

   a) Regional market b) Controlled market

   c) **Regulated market** d) Revised market

2. **The market charges are paid by the _____ of agricultural commodities in a regulated market.**

   a) **Buyers** b) Traders

   c) Farmers d) Sellers

3. **__________ is a voluntary business organization established by its members to market the farm products jointly to reap the direct benefit out of it.**

   a) Regulated market b) Regional market

   c) **Co-operatives** d) Collective market

4. **_____________ where its members are the owners, operators and contributors of the commodities marketed and they are the direct beneficiaries of the savings belongs to the society.**

   a) **Co-operative marketing society**

   b) Regulated marketing society

   c) Secular marketing society

   d) Direct marketing society

5. **Regulated marketing is supposed to reduce the marketing efficiency as it**

   a) Minimises the role of intermediaries

   b) Widens the price spread

c) Narrows the price spread

d) Adds less value

6. The co operative marketing society is controlled by _______.

a) Sellers
b) Buyers
c) Traders
d) Farmers

7. _____ acts as an insurance agent in Agricultural Co-operative Marketing.

a) APEDA
b) NAFED
c) APMC
d) NAIS

8. It is established by state governments where farmer can sell through open, competitive bidding with certain rules and regulations

a) Public Distribution System
b) Regulated markets
c) Co-operative markets
d) Farmers Mandi

9. _________ provides agricultural credit to the farmers for selling the produce immediately after harvesting.

a) APMC
b) Private Marketing Society
c) FCI
d) Co-operative marketing society

10. The market wherein, legislative measures are adopted to standardize the marketing function of agricultural produce with a aim to establish, upgrade and enforce standard marketing practices is __________

a) AGMARK
b) Regulated market
c) Cooperative Marketing
d) FCI

11. Which of the following does not match with regulated market

a) The legislation does not make it compulsory for the farmer to sell the produce

b) To establish, upgrade and enforce standardized marketing practices and charges

c) To market the commodity of the members in the society at fair prices

d) To promote a well-ordered marketing of agricultural produce by refining the infrastructural facilities.

12. __________ is referred as an association of producers for the collective marketing of their produce

a) Regulated Market **b) Cooperative Marketing**

c) Farmers Mandi d) AGMARK

13. NAFED stands for

a) National Agricultural Cooperative Marketing Federation

b) National Agricultural Federation

c) National Agricultural Federation for Development

d) National Agricultural Farmers Development

14. Under the Warehousing Act, it performances as warehouseman and construct its own godowns and cold storages

a) MAFED b) TRIFED

c) NAFED d) APEDA

15. Market intelligence is facilitated by

a) NABARD **b) Co-operatives**

c) Regulated markets d) Marketing Board

16. The most common method exist in regulated market is.

a) Moghum sale b) Close tender system

c) Open auction system d) Dara sale

17. Regulated marketing is supposed to reduce the marketing efficiency as it

a) Reduces the role of intermediaries

b) Adds less value

c) Widens the price spread

d) Reduces marketing cost

18. Regulated market guarantees

a) Minimum Support price b) **Fair price**

c) Procurement price d) Remunerative price

19. Regulated markets improve the efficiency of marketing by ______________

a) Ensuring proper weighment

b) Providing facilities like banking, insurance

c) Grading and standardizing

d) **All of these**

20. In unregulated market the charges taken for impurity is called as ______

a) Muddat
b) Darmoda
c) **Karda**
d) Dhelta

21. ______ provides agricultural credit to the farmers for selling immediately after harvesting the produce.

a) Private Marketing Society
b) FCI
c) **Co-operative marketing society**
d) NABARD

22. NAFED was established in co-operative sector at which of the following levels?

a) At national level
b) At state level
c) Specially for the farmers of tribal region of the country
d) Specially for the farmers of drought region of the country

23. ———— is an organization or set of organizations (go-betweens) involved in the process of making a product or service available for use or consumption by a consumer or business user.

a) Producer
b) Market intermediary
c) Farmer
d) None of the above

24. ———— is the function performed by the wholesalers in marketing.

a) Assemble the produce from various localities to meet the demands of buyers
b) Regulate the flow of produce by trading with buyers and sellers existing in the various markets
c) Balance the flow of produce by storing them in the peak season and releasing them in the off-season
d) All the above

25. The wholesalers perform which of the following functions?

a) Arrange the goods in different lots according to their quality and keep them for the market

b) Finance the farmers to meet their requirements of production inputs

c) Assess the demand of buyers and processors on time basis and prepare the movement of the goods over space and time

d) All the above

26. ———— are the producers personal representatives to the consumers.

a) Retailers b) Wholesalers

c) Facilitative middleman d) Commission agent

27. ———— is the function accomplished by the retailers in marketing.

a) Providing an assortment of products and services

b) Breaking bulk

c) Holding inventory

d) All the above

28. The retailers perform ______________ marketing functions

a) Provide marketing services

b) Provide product Information

c) Increase the value of products and services

d) All the above

29. ———— provide services that make it easier for customers to buy and use products.

a) Retailers b) Wholesalers

c) Producers d) Market brokers

30. Who display the products so consumers can see and test them before buying?

a) Wholesalers b) Producers

c) Retailers d) Market brokers

31. ———— purchase the produce of the farmers who have either get finance from them or those who are not able to go to the market and may supply essential consumption goods to the farmers.

a) Itinerant Traders b) Village Merchants

c) Retailers **d) Both a & b**

32. ———— sell only services to their market players and not the goods or commodities and also serve as buyers or sellers in effective bargaining.

a) Facilitative middleman **b) Agent middleman**

c) Retailer d) Wholesaler

33. ———— are the commission agents who mainly act for the sellers including farmers in the market.

a) Kaccha arhatias b) Pacca arhatias

c) Both a & b d) None of the above

34. ———— are the commission agents who act on behalf of the traders in the consumption side of market.

a) Kaccha arhatias **b) Pacca arhatias**

c) Both a & b d) None of the above

35. The processors and wholesalers in the consuming markets nominate ———— as their agents to purchase the specified quantity of goods.

a) Kaccha arhatias **b) Pacca arhatias**

c) Retailers d) Facilitative middlemen

36. In regulated markets, only one kind of commission agent who exists in the name of ———

a) Kaccha arhatias b) Pacca arhatias

c) 'A' class trader d) None of the above

37. Commission agents extend ———— facilities to their clients.

a) Advance 40 % to 50 % of the projected value of the crop as a loan to the farmers

b) Provide storage facilities and advance loans to the farmers against the stored commodities up to 75 % of its value

c) Help the farmers during their personal difficulties

d) All the above

38. ———— combine buyers and sellers on the same stand for negotiations but do not have any physical control over the product.

a) Brokers b) Commission agents

c) Wholesalers d) Retailers

39. ———— are the characteristics of brokers in an agricultural market.

a) No establishment in the market

b) Do not take any risk

c) Do not render any other service except to bringing together the buyers and sellers to the same point

d) All the above

40. ———— is a facilitative middleman.

a) Labourers b) Weighmen

c) Graders **d) All the above**

41. Transport Agency, Communication Agency and Advertising Agency belong to ____________

a) Facilitative middlemen b) Agent middlemen

c) Speculative middlemen d) Merchant middlemen

42. In the case of sugarcane, spices and oilseeds which require processing before consumption, the marketable surplus as a portion of total output is ———— than that of other food crops.

a) Smaller **b) Larger**

c) Negligible d) Insignificant

43. The inverse relationship between marketed surplus and price level depicts that farmers have ———— cash requirements.

a) Inelastic b) Elastic

c) Minimum d) High

44. The positive relationship between marketed surplus and prices of food grains assumes that farmers are price ————

a) Takers b) Makers

c) Conscious d) None of the above

45. According to Rajkrishna, the elasticity of the marketable surplus is ———— so long as the substitution effect is non-zero.

a) Negative **b) Not negative**

c) Not affected d) None of the above

46. ———— model indirectly investigates the size and magnitude of the elasticity of the marketed surplus of a subsistence crop.

a) Rajkrishna Models b) Behrman Model

c) T.N. Krishnan Model **d) All the above**

47. The state of selling the commodity more than marketable surplus is termed as ————

a) Profitable sale b) Wholesale

c) Conspicuous sale **d) Distress sale**

48. When cash need of small and marginal farmers is immediate, marketed surplus is ———— the marketable surplus.

a) Less than **b) More than**

c) Equal to d) None of the above

49. Large farmers usually sell ———— than the marketable surplus because of their better retention capacity.

a) Less b) More

c) Same d) None of the above

50. In the case of perishable commodities and agricultural raw materials like cotton, jute, marketed surplus is ———— the marketable surplus.

a) More than b) Less than

c) Equal to d) None of the above

51. Any business firm that operates between producers and consumers is generally referred as ————

a) Marketing Intermediary b) Wholesaler

c) Retailer d) None

52. ———— is a marketing Intermediary

a) Facilitating intermediary b) Wholesaler

c) Retailer **d) All the above**

53. The middlemen who involved in marketing of food grains are classified as

a) Merchant middle men b) Agent middlemen

c) Speculative middlemen d) All the above

54. ———— are the middlemen who take title to the goods they handle.

a) Merchant middlemen b) Agent middlemen

c) Speculative middlemen d) Facilitative middlemen

55. Wholesalers are ————

a) Merchant middlemen b) Agent middlemen

c) Speculative middlemen d) Facilitative middlemen

56. ———— buy produce as a whole in large quantities and sell them to the consumers in small quantities.

a) Retailers b) Wholesalers

c) Agent middlemen d) Speculative middlemen

57. ———— are simple petty merchants who move from one village to another village and directly purchase the produce from the farmers.

a) Itinerant traders b) Retailers

c) Wholesalers d) Agent middlemen

58. ———— act as representatives of their clients and do not take any title to the produce and therefore, do not own the produce too.

a) Merchant middlemen **b) Agent middlemen**

c) Speculative middlemen d) Facilitative middlemen

59. ———— are agent middlemen

a) Commission agents b) Brokers

c) Both a & b d) None

60. The middlemen who take title to the produce with aim of making a profit from it are called ————

a) Merchant middlemen b) Agent middlemen

c) **Speculative middlemen** d) Facilitative middlemen

61. ———— do not directly buy and sell the commodity but helps in the marketing process and increase the efficiency and marketing can take place even if they are not available

a) Merchant middlemen b) Agent middlemen

c) Speculative middlemen **d) Facilitative middlemen**

62. ———— is a person involving in the wholesale market acting as the representative of a seller or a buyer.

a) Merchant middlemen b) **Commission agent**

c) Broker d) None

63. ———— render their personal services to their clients in the market but do not have any physical control over the product.

a) Merchant middlemen b) Commission agent

c) **Broker** d) None

64. ————————————, if efficiently organized, can help reduce the price-spread between the producer and the consumer, thereby assisting in giving a fair return to the producer without adversely affecting the legitimate interests of the consumer

a) Marketing **b) Cooperative marketing**

c) Market integration d) Efficient marketing

65. ——————————— is a process of establishing a voluntary business organization by its member patrons to market the farm products together for their direct benefit from it.

a) Collaboration b) Conglomeration

c) Cooperation d) None of the above

66. ———is not the functions provided by the co-operative marketing societies

a) Assurance of fair prices

b) Safeguard the members from excessive marketing costs and malpractices

c) Non - credit facilities

d) None of the above

67. The cooperative marketing societies have

a) Two tier b) Three tier

c) Both d) Four Tier

68. Members in co-operative marketing societies

a) Ordinary b) Nominated

c) Real **d) Both a and c**

69. Sources of Finance for co-operative marketing societies

a) Share of capital b) Loans

c) Subsidy **d) All the above**

70. COSAMB

a) Council of State Agricultural Marketing Boards

b) Council of State Agricultural Marketing Bureau

c) Centre for State Agricultural Marketing Boards

d) Centre for State Agricultural Marketing Bureau

71. The COSAMB, an apex body of the State Marketing Boards was established in

a) 1988 b) 1987

c) 1889 d) 1897

72. Markets where business is done in harmony with the rules and regulations framed by the statutory organization is called as ____________

a) Secondary Market **b) Regulated market**

c) Primary market d) None of the above

73. The specific objectives of the regulated markets are

a) To prevent the handicaps in the marketing of their products,

b) To make the farmers to get better prices for their produce and the goods are made available to consumers at reasonable prices

c) To provide incentive prices to farmers for inducing them to increase the production

d) All the above

74. Cooperative marketing is not successful in India except

a) Milk b) Cotton

c) Edible oil d) Pulses

75. In regulated markets, ____ are regulated

a) Prices b) Farmers behaviour

c) Traders behaviour **d) Marketing practices**

76. The regulated market have been established by ________

a) Central government b) District Municipal Corporation

c) State government d) Panchayats

77. DMI stands

a) Department of Marketing and Inspection

b) Directorate of Marketing and Inspection

c) Directorate of Marketing and Information

d) Department of Marketing and Information

78. NCAER stands

a) National Council of Applied Economic Region

b) National Centre for Applied Economic Region

c) National Council of Applied Economic Research

d) National Centre for Applied Economic Research

79. Directorate of Marketing and Inspection (DMI) started at

a) 1934 **b) 1935**

c) 1953 d) 1943

80. ———— is an initiative of ITC Limited, a large multi business conglomerate in India, to link directly with rural farmers via the Internet for procurement of agricultural and aquaculture products

a) e-Choupal b) AGMARKNET

c) eBAY d) DMI

81. __________ has been compared to a "State within A State"

a) State **b) Cooperation**

c) Capitalism d) Socialism

82. __________ plays an equalizing role as a welfare factor in a capitalistic economy.

a) Socialism b) Trade Unionism

c) Joint Stock Companies **d) Cooperation**

83. Regulated markets are markets where business is conducted

a) Without any set of rules and regulations

b) With the rules and regulations framed by statutory market

c) With Bonds, shares and securities are bought and sold

d) None of the above

83. In India, ____________________ is regulated by the Forward Market Commission?

a) Currency future trading **b) Commodity future trading**

c) Equity future trading d) Both a and b

84. ———— is the prime regulator of commodity futures markets in the country

a) Forward markets commission

b) NAFED

c) Commodity trade association

d) None of the above

4

Warehousing and Quality Control of Agricultural Products

1. ———— **is the study of all the activities, agencies and policies involved in the procurement of farm inputs by the farmers and the movement of agricultural products from the farms to the consumers.**

   a) Agricultural marketing b) Supply chain

   c) Agricultural production d) Value chain

2. **The subject of agricultural marketing comprises of ————**

   a) Product marketing b) Input marketing

   c) Both a & b c) None of the above

3. **Changing the raw material into a finished good creates ————**

   a) Form utility b) Time utility

   c) Place utility d) Possession utility

4. **Form utility is created by the ————**

   a) Processing function b) Transportation function

   c) Storage function d) Marketing function

5. ____________ **is an example for creating form utility**

   a) Processing of oilseeds into oil

   b) Preparation of flour and bread from wheat

   c) Sugar form sugarcane

   d) All the above

6. **Shifting the goods from a place of plenty to a place of need creates ————**

   a) Form utility b) Time utility

   c) Place utility d) Possession utility

7. Place utility is created by the ———

a) Processing function **b) Transportation function**

c) Storage function d) Marketing function

8. Storage function adds ——— to the products

a) Form utility **b) Time utility**

c) Place utility d) Possession utility

9. Transfer of ownership from one person to another creates____

a) Form utility b) Time utility

c) Place utility **d) Possession utility**

10. Marketing activities of buying and selling creates ———

a) Form utility b) Time utility

c) Place utility **d) Possession utility**

11. Which of the following is a characteristic of agricultural production?

a) Small Scale Production

b) Scattered and Specialized Production

c) Seasonal Production

d) All the above

12. The dependency of agricultural production on natural factors and conditions makes the supply of agricultural products ———

a) Uncertain and Irregular b) Risk free

c) Both a & b d) None

13. The features of agricultural commodities include ———

a) Variation in quantity b) Variation in quality

c) Perishable **d) All the above**

14. The demand for most agricultural products is ———

a) Relatively inelastic b) Relatively elastic

c) Unitary elastic d) Perfectly elastic

15. The adjustment of supply to demand for agricultural commodities takes place ———

a) Rapidly **b) Slowly**

c) Quick pace d) None

16. The prices of agricultural products fluctuate significantly due to ____________

a) Irregular supply b) Constant demand

c) Both a & b d) None

17. Compared to manufactured goods, the marketing of agricultural products tends to be a _________

a) Complex process b) Simple process

c) Less risky d) None

18. One of the major physical characteristics of agricultural production is _________

a) Bulky production b) Perishable nature

c) Both a & b d) None

19. The processing function increases _________ of agricultural commodities

a) Price spread b) Shelf-life

c) Both a & b d) None

20. The major problem with bumper harvest of agricultural commodities is ___________

a) Distress sale b) High income

c) Reduced demand d) None

21. __________ is an important marketing function, by which goods are detained and preserved from the time they are produced until disposed for consumption.

a) Storage b) Transportation

c) Processing d) None of the above

22. __________ is the process of transferring goods from one place to another

a) Storage **b) Transportation**

c) Processing d) None of the above

23. Need and importance of Storage in marketing are _________

a) To protects the quality of the perishable and semi-perishable products from deterioration

b) To stabilization of price by adjusting demand and supply

c) To provides income and employment through price advantage

d) All the above

24. ———— are the risk in storage of agricultural commodities.

a) Quantity loss from rodents, insect and pest, theft and fire

b) Quality deterioration from pest, excessive moisture and temperature.

c) Price risk from fall in price

d) All the above

25. Increase in the price of the stored product at the time of storage till it is dispatched is ———— on storage.

a) Gross return b) Net return

c) Gross cost d) Net cost

26. The cost of storage should include ————

a) The cost of the maintenance of the storage structure and the cost of protective materials

b) Tax payments and payments to labour and interest on the value of the stored goods

c) Value of the quantitative and qualitative loss during storage and risk premium for a possible price fall and damage

d) All the above

27. Net returns to storage is calculated as ————

a) NR = GR – C b) NR = GR + C

c) NR = GR / C d) NR = GR * C

28. The percentage margin (Ms) from product storage is calculated as ————

a) **$Ms = (P_1 - P_0 - C) / (P_0 + C)$** b) $Ms = (P_1 - P_0 / C) - (P_0 + C)$

c) $Ms = (P_1 - P_0 + C) / (P_0 - C)$ d) None of the above

29. The AGMARKNET portal provides information on

a) Prices and arrivals

b) Trend analysis of prices and arrivals

c) Grading, packaging and marketing charges

d) All the above

30. The entire system of and the name of AGMARK was created by

a) Livingstone b) Lemark

c) J. S. Chaddha d) Johnston

31. Which one is not a certification mark in India for food products?

a) FSSAI b) FPO

c) AGMARK d) India Organic

32. Central Agmark Laboratory is located at

a) Pune **b) Nagpur**

c) Kanpur d) Patna

33. Promotion of grading and standardization of agricultural and allied products under the Agricultural Produce (Grading & Marketing) Act. 1937

a) DMI b) NIAM

c) MANAGE d) AMI

34. The Agricultural Produce (Grading & Marking) Act,1937 empowers the Central Government to fix quality standards known as

a) FCI b) NIAM

c) AGMARK d) DMI

35. Agricultural Produce (Grading & Marketing) Act.

a) 1937 b) 1935

c) 1957 d) 1953

36. Meat Food Products Order

a) 1953 **b) 1973**

c) 1983 d) 1963

37. Essential Commodities Act

a) 1956 b) **1955**

c) 1945 d) 1965

38. International food safety standards are developed by

a) HACCP **b) CODEX**

c) ECOMARK d) WTO

39. Find out the true statement regarding to AGMARK

a) AGMARK is for processed products

b) It is a compulsory scheme

c) **Certification under the Agricultural Produce act**

d) Label colour indicates the variety of the product

40. "AGMARK replica" means

a) Grade designation mark	b) Product designation mark
c) Variety designation mark	d) Brand designation mark

41. An integrated scheme of scientific storage, price stabilization, rural credit and market intelligence

a) Warehouses	b) Co-operatives
c) FCI	d) NABARD

42. These warehouses are specially constructed at a seaport or an airport and accept imported goods for storage till the payment by the importer.

a) Smart Warehouses	b) Special Commodity Warehouses
c) General Warehouses	**d) Bonded warehouses**

43. The establishment of the Central Warehousing Corporation was established on

a) 1973	b) 1947
c) 1957	d) 1937

44. Nationalized banks provide credit on the security of warehouse receipt of the stored products to the extent of

a) 75 % of their value	b) 65 % of their value
c) 63 % of their value	d) 73 % of their value

45. National Co-operative Development and Warehousing board was established by

a) 1965	b) 1976
c) 1966	**d) 1956**

46. Warehousing Corporation Act

a) 1957	**b) 1962**
c) 1952	d) 1956

47. __________ is a part of development of facility structures.

a) Transportation	**b) Warehousing**
c) Sorting	d) Logistics

48. FPO mark certificate is given to ......

a) Agricultural products	b) Industrial products
c) Eco friendly products	**d) All processed food products**

49. Which agency is responsible for procurement, distribution and storage of food grain production in India?

a) Ministry of Agriculture **b) Food Corporation of India**

c) NAFED d) TRIFED

50. In India, which one among the following is not having certification mark for food products?

a) FSSAI **b) AGMARK**

c) FPO d) HACCP

51. The Agricultural Produce (Grading & Marking) Act,1937 authorizes the Central Government to fix quality standards known as

a) FCI b) NIAM

c) AGMARK d) DMI

52. Number of notified agricultural and allied commodities under AGMARK

a) 88 b) 133

c) 506 d) **227**

53. National Institute of Agricultural Marketing has started from

a) 1986 b) 1989

c) **1988** d) 1983

54. Cold Storage Order

a) **1980** b) 1986

c) 1968 d) 1960

55. An integrated scheme of scientific storage, price stabilization, rural credit and market intelligence

a) Warehouses b) Co-operatives

c) FCI d) NABARD

56. These warehouses are specially constructed at a seaport or an airport and accept imported goods for storage till the payment by the importer.

a) Smart Warehouses b) Special Commodity Warehouses

c) General Warehouses d) **Bonded warehouses**

57. The establishment of the Central Warehousing Corporation was in

a) 1973 b) 1947

c) 1957 d) 1937

58. The establishment of a National Co-operative Development and Warehousing board was on

a) 1965 b) 1976

c) 1966 **d) 1956**

59. Food Corporation of India was setup on

a) 1965 b) 1955

c) 1956 d) 1966

60. Foodgrains distribution by Public Distribution System was throughout the country is done by

a) ITC b) AGMARK

c) FCI d) HACCP

61. Sustaining satisfactory level of buffer stocks of foodgrains to ensure National Food Security is undertaken by

a) ITC b) NFSM

c) FCI d) APEDA

63 ________ is part of development of facility structures.

a) Transportation **b) Warehousing**

c) Sorting d) Logistics

64. FPO mark certificate is given to ___________

a) Agricultural products b) Industrial products

c) Eco friendly products **d) All processed food products**

65. APEDA is under the Ministry of ________.

a) Agriculture and Farmers Welfare

b) Commerce and Industry

c) Consumer affairs, food and public distribution

d) Ministry of Food Processing Industries

66. Quality certification marks under the Central Agricultural Produce (Grading and Marketing) Act, 1937.

a) ECO MARK b) **AGMARK**

c) FPO d) BIS

67. International food safety standards are developed by

a) HACCP **b) CODEX**

c) ECOMARK d) WTO

68. Which one is not a certification mark in India for food products?

a) FSSAI
b) FPO
c) AGMARK
d) India Organic

69. Standards for agricultural grading have been fixed by ——————————

a) Agricultural Marketing Advisor, Government of India
b) Food Safety and Standards Authority of India
c) The Deputy Director General, Bureau of Indian Standards.
d) Directorate of Standardisation and Quality Certification

70. Risks associated with storage

a) Price Risk
b) Financial risk
c) Technological risk
d) Weather risk

71. The abnormal increase in buffer stock of FCI in recent times is due to

a) Procurement prices are above market prices
b) Additional storage facility created by FCI
c) Transport bottlenecks in wheat producing regions
d) Series of bumper harvest

72. Products are transferred through marketing from persons having a ____________ utility to persons having a ___________ utility

a) Higher; Lower
b) Lower; Higher
c) Same level of utility
d) None of the above

73. Products command higher prices at the place of necessity than that of the place of production because of the ———— utility of the product.

a) Lower
b) Higher
c) Same level of utility
d) None of the above

74. Agricultural production in India is characterized by ————

a) Large-scale production
b) Small-scale production
c) Commercial production
d) Industrial production

75. From year to year, amount and quality of agricultural products — ———

a) Varies b) Remains same

c) Unaffected d) In-variable

76. Price fluctuation is ——— in agricultural commodities.

a) Rare b) Not possible

c) Common d) Minimum

77. Bulky nature of agricultural products makes transportation and storage ———

a) Easy **b) Difficult**

c) Affordable d) Smooth

78. Processing function ——— the price spread of agricultural commodities.

a) Increases b) Decreases

c) Not affect d) Not influence

79. ——— is necessary for the performance of marketing functions till arrangement for transport, during the process of buying and selling.

a) Storage b) Processing

c) Packing d) None of the above

80. ___ protects mainly the quality of perishable and semi-perishable products from deterioration.

a) Transportation **b) Storage**

c) Packing d) Standardization

81. ——— vary with the quantity of goods stored.

a) Fixed storage cost **b) Variable storage cost**

c) Overhead cost d) None of the above

82. ——— does not vary with the quantity of goods stored.

a) Fixed storage cost b) Variable storage cost

c) Marginal storage cost d) None of the above

5

Producer Surplus Marketed Surplus and Marketable Surplus

1. **At times of distress sales**

    a) Marketed surplus < Marketable surplus

    b) Marketed surplus > Marketable surplus

    c) Marketed surplus = Marketable surplus

    d) Zero marketed surplus

2. **For durable goods, under normal pricing conditions**

    a) Marketed surplus < Marketable surplus

    b) Marketed surplus > Marketable surplus

    c) Marketed surplus = Marketable surplus

    d) Zero marketed surplus

3. **__________ is the quantity of the produce which can be made available to the non-farm population of the country after meeting farmers' family consumption, farm requirements, social and religions payments.**

    a) Marketable surplus b) Marketed surplus

    c) Both a & b d) None

4. **Marketable surplus may be expressed as ————**

    a) MS = Total Production – Total requirement of farm family

    b) MS= Total production / Total requirement of farm family

    c) MS= Total requirement of farm family / Total production

    d) None

5. **Marketable surplus depends on ————**

    a) Size of Holding b) Production

    c) Size of family **d) All the above**

6. **The price of the commodity and market-able surplus has a __________ relationship.**

a) Positive b) Negative

c) Both a & b d) No relationship

7. **There is higher production in the farm, the ———— will be the marketable surplus**

a) Larger b) Smaller

c) Not affected d) None of the above

8. **Marketable surplus of non-food crops is generally ———— than that of food crops**

a) Higher b) Lower

c) Same d) None of the above

9. **According to P.N. Mathur and M. Ezekiel, there is ———— relationship between the marketable surplus and prices.**

a) Direct **b) Inverse**

c) No relationship d) None of the above

10. **According to V.M. Dandekar and Rajkrishna, there is ———— relationship between prices and the marketed surplus of food grains in India.**

a) Negative **b) Positive**

c) Inverse d) No relationship

11. **———— is that quantity of the produce, which the farmer actually sells in the market, irrespective of his requirements for family consumption, farm requirements, social and religious payments.**

a) Marketable surplus **b) Marketed surplus**

c) Both a & b d) None

12. **The marketed surplus may be ———— than/to the marketable surplus.**

a) More b) Less

c) Equal **d) All the above**

13. **The marketed surplus is ———— than the marketable surplus when the farmer retains a smaller quantity of crop than his actual family and farm requirements.**

a) More b) Less

c) Equal d) None of the above

14. The marketed surplus may be ———— the marketable surplus when the farmer neither retains more nor less than his requirement.

a) More b) Less

c) **Equal** d) None of the above

15. Most of the farmers have limited access to the market information due to ———— literacy level.

a) Low

b) High

c) Access to information is not related to literacy level

d) None of the above

16. Producers share in consumers rupee ———— with ———— in number of intermediaries.

a) Increases; increase **b) Increases; decrease**

c) Decreases; decrease d) Not related

17. If a product is in short supply, its price is likely to ————

a) Increase rapidly b) Increase gradually

c) Decrease rapidly d) Decrease gradually

6

Product Life Cycle (PLC)

1. **What does the term PLC stands for?**

   a) Product long cycle
   b) Production life cycle
   c) **Product life cycle**
   d) Production long cycle

2. **Which of the following is not the stage of Product Life Cycle?**

   a) Introduction Stage
   b) Peak stage
   c) Decline stage
   d) Mature stage

3. **When a new product enter into the market with higher quality and new features better than its competitors, those products are known as__________**

   a) Superior products
   b) Develop superior products
   c) New products
   d) **Unique superior products**

4. **Which of the following is not a characteristic of "Market Introduction Stage" in PLC?**

   a) There is no competition
   b) Low cost
   c) Makes no revenue
   d) Slow sales volume to start

5. **"A product lifecycle is very much similar to human life cycle" stated by ______**

   a) **Philip Kotler**
   b) Stanton
   c) Neil Borden
   d) **Arch Paton**

6. **Increased competition, price decrease, increasing public awareness, increased sales volume are the characteristics of ______________ in PLC.**

   a) Mature stage
   b) Decline stage
   c) Growth stage
   d) Market introduction stage

7. ____________________ **includes review of sales, profit projections and cost for a new product, to find out whether it satisfied the company objective or not.**

a) Product Development b) Marketing Strategy

c) Business Analysis d) Idea Screening

8. **The most important performance dimension for product development project is**

a) Time to market b) Time to target

c) Time to consumer d) Time to seller

9. **Type of advertisement is used when the product enters into growth stage of PLC**

a) Primary advertising b) Reminder advertising

c) **Selective advertising** d) None of these

10. **The strategy of choosing one attribute to excel to create competitive advantage is known as**

a) Under positioning b) **Unique selling proposition**

c) Over positioning d) None of these

11. **Color and size of the product, brand and packaging are considered as**

a) Chemical features of product b) **Physical features of product**

c) Product designing d) Product manufacture

12. ________________________ **is one of the challenges imposed by the Product Life Cycle.**

a) Product development **b) New product development**

c) Product testing d) Poor margins

13. **In "Product Life Cycle" a stage represents rapid growth of product sale knows as**

a) Market introduction phase b) Mature phase

c) Saturation phase d) **Growth phase**

14. **Rising profits is a feature of___________ stage of PLC.**

a) Growth b) Introduction

c) Maturity d) Saturation

15. Revival plans to reintroduce the product in more modified form is adopted in ___________stage of PLC.

a) Introduction b) Maturity

c) Decline d) Growth

16. Launch better advertising campaign and use aggressive sales promotions like trade-deals, discounts, premiums, contests etc is

a) Modifying the product b) Modifying the market

c) Modifying the market mix d) Modifying the process

17. The products enters maturity when

a) Increase in sale

b) Decrease in profit

c) Sales start growing

d) Sales stop growing and demand stabilises

18. Who developed SWOT analysis?

a) Scot Hall b) **Albert Humphrey**

c) Robert Plant d) James Hannma

19. Who developed the concept of marketing mix?

a) Abraham Maslow b) Stanton

c) **N.H Borden** d) Philip Kotler

20. What is the aim of marketing process?

a) Production b) Profit making

c) **Satisfaction of customer needs** d) Selling products

21. Marketing mix is made of __________.

a) Production recognition b) Price structure

c) Distribution planning d) **All of above**

22. Price, product, promotion and place forms a group of

a) Market mix b) **Marketing mix**

c) Product mix d) Promotion mix

23. The only revenue producing element in marketing mix is

a) Product b) **Price**

c) Place d) Promotion

24. This P is not a part of the 7Ps of marketing mix?

a) Promotion b) People

c) **Purpose** d) Pricing

25. Which is not the method of cost based pricing?

a) Cost plus pricing b) Marginal cost pricing

c) **Differential pricing** d) Target pricing

26. Methods of competition based pricing

a) Going rate pricing b) Sealed bid pricing

c) Customary pricing d) **All of these**

27. When there is large potential market for product the firm will adopt

a) Skimming price b) Penetration price

c) **Premium price** d) None of these

28. A market may be segmented by classifying people according to their enthusiasm for a product are termed as

a) Attitude segmentation b) Geographic segmentation

c) Socio-economic segmentation d) Psychographic segmentation

29. Which is not the stage of product life cycle?

a) Sales decline b) Market maturity

c) Market introduction **d) Market implementation**

30. The E-commerce domain that involves business activity initiated by the consumer and targeted to businesses is known as

a) B2B b) C2C

c) B2C d) C2B

31. The advertisement must be genuine and accurate creates

a) Conviction value b) Attention value

c) Instinctive value d) Educative value

32. The stage is the product life cycle that focuses on expanding market and creating product awareness and trial is the

a) Decline stage
b) Introduction stage
c) Growth stage
d) Maturity stage

33. The object of sales promotion is to increase the buying response of ultimate

a) Consumer
b) Wholesaler
c) Retailers
d) Manufacturers

34. __________ is a marketing process which aims at the collection of products at a central place.

a) Concentration
b) Dispersion
c) Processing
d) Equalization

35. The products meant for ultimate users are subdivided into small lots and distributed to meet the final consumption.

a) Concentration
b) Equalization
c) Dispersion
d) Assembling

36. It implies the reconciliation between demand and supply through storage and transportation.

a) Equalization
b) Concentration
c) Assembling
d) Dispersion

37. Advertising creates _____ among customers.

a) Clarity
b) Awareness
c) Confusion
d) Belief

38. Promotional mix includes _____.

a) Advertising, awareness and sales promotion.
b) Advertising, personal selling and publicity.
c) Advertising, personal selling and sales promotion.
d) Segmentation, personal selling and sales promotion.

39. The advertisement must be genuine and accurate creates ——— —— values

a) Conviction value
b) Attention value
c) Instinctive value
d) Educative value

40. _________ is an online advertisement that pops up between changes on a website

a) Border **b) Plunge**

c) Boarder d) Interstitial

41. Informing buyers of new brand and new package is the________of sales promotion.

a) Objective b) Method

c) Application d) Quality

42. The essential criteria for effective segmentation is

a) Homogenity b) Measurability

c) Profitability **d) All of these**

43. Identifying and providing different marketing mix for each of the segments is known as___________

a) Undifferentiated Marketing **b) Differentiated Marketing**

c) Concentrated Marketing d) Customised or Personalised Marketing

44. If the product passes through a longer channel of distribution, the marketer will have to give importance to

a) Advertising b) Personal selling

c) Direct selling d) None of these

45. If advertising give focus on a particular product or brand, it is known as

a) **Product advertisement** b) Trade advertisement

c) Institutional advertisement d) Market advertisement

46. When the advertisement is to create an image or reputation of the firm, it is a case of

a) Product advertisement b) **Institutional advertisement**

c) Reminder advertising d) None of these

47. Advocacy advertising is also called

a) Support advertising b) Rapport advertising

c) **Cause advertising** d) None of these

48. A good catchy phrase used and repeated often in an advertisement is ———

a) Idea b) Brand name

c) Trade mark **d) Slogans**

49. ____________ is not a commercial advertising

a) Consumer ad b) Industrial ad

c) Trade ad d) **Shortage ad**

7

Marketing Channel Marketing Cost and Price Spread

1. **Price spread will be the minimum in**

   a) Coconut b) Rice

   c) Milk **d) Green leaves**

2. **Farmers share in consumer rupee will be the minimum in**

   a) Gram b) Milk

   c) Cotton d) Rice

3. **Marketable surplus is equal to**

   a) Total Consumption –Total Production

   b) Total Production

   c) Total Consumption

   d) Total Production – Total Consumption

4. **Price spread includes**

   a) Marketing cost+ Market margin

   b) Marketing cost+ Profit margin

   c) Marketing cost+ Concurrent margin

   d) None of the above

5. **The expenditure incurred farmers & various market intermediaries in transacting the produce is called**

   a) Marketing cost b) Profit margin

   c) Concurrent margin d) Price spread

6. **In general, for large farmers, marketable surplus is ——— marketed surplus**

   a) Equal to **b) Less than**

   c) More than d) None of the above

7. **In general, for small farmers, marketable surplus is ——— marketed surplus**

a) Equal to b) Less than

c) More than d) None of the above

8. **For perishable commodities, marketable surplus is ——— marketed surplus**

a) Equal to b) Less than

c) More than d) None of the above

9. **__________ method of sale offers more competitive price to the farmers**

a) Closed tender b) Hatha

c) Open auction d) Dara

10. **In ___________ market, future buying and selling of commodities will take place at current time**

a) Spot **b) Forward**

c) International d) Perfect

11. **At times of distress sales**

a) Marketed surplus < Marketable surplus

b) Marketed surplus > Marketable surplus

c) Marketed surplus = Marketable surplus

d) Zero marketed surplus

12. **For durable goods, under normal pricing conditions**

a) Marketed surplus < Marketable surplus

b) Marketed surplus > Marketable surplus

c) Marketed surplus = Marketable surplus

d) Zero marketed surplus

13. **The marketing cost of processed product is _________ than raw product**

a) Lower **b) Higher**

c) Equal d) None

14. The marketing cost of goods sold in lean season is _________ than in peak season

a) Lower **b) Higher**

c) Equal d) None

15. __________ refers to the price differences between successive stages of marketing at a given point of time

a) Lagged margin b) Marketing cost

c) Price spread **d) Concurrent margin**

16. Price differences between farmers selling price and retail price on the same date refers to

a) Lagged margin b) Marketing cost

c) Price spread d) Concurrent margin

17. Which of the following takes into consideration the time that elapses between purchase and sale of commodity in the market?

a) Lagged margin b) Marketing cost

c) Price spread d) Concurrent margin

18. Price fixed by the government to protect farmers against excessive fall in prices

a) Floor price b) Procurement price

c) Issue price **d) MSP**

19. ________ refers to the price at which government procures from producers to maintain buffer stocks and feed Public Distribution System

a) Floor price **b) Procurement price**

c) Issue price d) MSP

20. Price at which the commodity is made available to consumers at fair price shops. It is always higher than procurement price

a) Floor price b) Procurement price

c) Issue price d) MSP

21. Middlemen will increase the _______________

a) Price of the product

b) Quality of the product

c) Profit of the product

d) Time and place utility of the product

22. Price is a ________ term

a) Absolute b) Relative

c) Composite d) Standard

23. ________ is price at which a retailer sells the products to his buyers

a) Retail price b) Whole sale price

c) FOB price d) Administered price

24. Products reach the hands of customers through a number of channels, of that the main channel is____________

a) Wholesaler b) Distributor

c) Retailer d) Agents

25. Selling is an act of ________

a) Persuasion b) Illusion

c) Forcing d) Communication

26. Calculating marketing margin and marketing cost for fresh fruits involves __________

a) Lot method

b) Comparison of prices at successive levels of marketing

c) Sum of average gross margin method

d) Both b and c

27. In marketing intermediaries, the way of distribution in which the product is stocked in many possible outlets is classified as

a) Inclusive distribution b) Exclusive dealing

c) Selective distribution **d) Intensive distribution**

28. If the cost of product is $30 and the profit margin for each unit is $3 then the price must be charged to customer is

a) $30 **b) $33**

c) $27 d) $34

29. The kind of channel arrangement which involves one or more than one independent wholesalers, producers and retailers is classified as

a) Vertical marketing system

b) Static distribution channel

c) Conventional distribution channel

d) Horizontal marketing system

30. The most crucial and first step in marketing process is

a) Designing a market strategy

b) Create customer delight

c) Understanding customer needs and wants

d) Capturing value from customers

31. The channel alternative is not to be assessed on the basis of

a) Economic criteria

b) Control criteria

c) Adaptive criteria

d) Accumulation criteria

32. The retail supply chain does not include

a) Manufactures

b) Retailers

c) Wholesalers

d) Regulators

33. The types of good value pricing includes

a) Everyday low pricing

b) High low pricing

c) Low high pricing

d) Both a and b

34. A marketing channel is

a) A set of interdependent organizations involved in the process of making a product or service available for use or consumption

b) A manner by which products or services are sold

c) A method for communicating advertising

d) A process by which customer feedback can be provided to the manufacturer

35. Four factors that create change in marketing channels are;

a) Facilitation of Search, Routinization of Transactions, Adjustment of Assortment Discrepancy, and the Reduction of the number of contacts

b) Environment, Government, Financials, and Competition

c) Demographics, Psychographics, Politics, and Market Size

d) Environment, Government, Politics, and Competition

36. The work of a marketing channel includes the performance of___________

a) Several marketing flows

b) Primarily, keeping customers satisfied

c) Making products

d) Keeping inventory levels low

37. Marketing channels consist of

a) Manufacturers, Intermediaries, and End-Users

b) Manufacturers, Wholesalers, and Retailers

c) Service Providers, Manufacturers, and Advertisers

d) Trade barges navigating coastal waters

38. A _________ of distribution is the route taken by the tittle to the product as it moves from the producer to the ultimate consumer or industrial user

a) Highway b) Canal

c) Channel d) Place

39. A ________ level channel is one in which there are no intermediaries

a) Zero b) One

c) Two d) Three

40. Marketing channels form a sub variable of ________ mix

a) Place b) Price

c) Promotion d) Product

41. Marketing ——————— creates time, place and possession utilities

a) Plan b) Department

c) Channels d) Idea

42. ____________ are more common in industrial products and high priced complex products like computers

a) Direct channels b) Indirect channels

c) Forward channels d) Backward channels

43. —————— refers to distribution of goods with the help of intermarries

a) Direct channels **b) Indirect channels**

c) Forward channels d) Backward channels

51. ————— efficiency either relates to functional deficiencies or to the degree of competition or monopoly and to economic structure existing within the marketing system.

a) Operational Efficiency **b) Pricing Efficiency**

c) Physical Efficiency d) None of the above

52. ———— criteria may be used to measure marketing efficiency.

a) Price spread
b) Market integration
c) Both a & b
d) None of the above

53. ———— is a type of e-marketing company.

a) Pure-click companies
b) Business to Business E-commerce
c) Brick and Click companies
d) All the above

54. ———— refers to structural characteristics of marketing system where sellers are able to get true value of their produce and consumers receive true worth of money.

a) Operational Efficiency
b) Pricing Efficiency
c) Physical Efficiency
d) None of the above

55. ———— bridges the gap between the producers and users.

a) Market structure
b) Market conduct
c) Market integration
d) Marketing channel

56. The channel of distribution can be classified into ————

i) Direct channels
ii) Indirect channels
iii) Service Sector Channels
iv) Channel Integration and Systems
v) E- Marketing channels

a) (i) & (ii)
b) (i) (ii) (v)
c) (i) (ii) (iii) (v)
d) (i) (ii) (iii) (iv) (v)

57. ———— is the oldest, shorter and the simple channel of distribution.

a) Direct channels
b) Indirect channels
c) Service Sector Channels
d) E- Marketing channels

58. Which of the following is an example for direct marketing channel?

a) Direct mail
b) Catalog sales
c) Telemarketing
d) All the above

59. ____________ is not correct regarding direct marketing channel?

a) Availability of expert services of middle man

b) Small investment

c) Suitable for small producers

d) All the above

60. _____________ are the methods of selling through direct marketing channel?

a) Own retail outlets

b) Postal services and courier services

c) Telemarketing

d) All the above

61. ________________ goods are suitable for selling through direct marketing channel?

a) Costly industrial goods

b) Perishable goods

c) Household appliances

d) All the above

62. ________ is known as traditional or normal channel of distribution in case of indirect marketing channel.

a) Producers -> Wholesalers -> Retailers -> Customer (Two level Channel)

b) Producer -> Agent -> Retailer -> Consumer (Two level Channel)

c) Producer -> Retailer -> Customer (One level Channel)

d) None of the above

63. ———— comes under Channel Integration and Systems.

a) Vertical Marketing System

b) Horizontal Marketing systems

c) Service Sector Channels

d) a & b only

64. ———— is a form of Vertical Marketing System.

a) Corporate

b) Administered

c) Contractual

d) All the above

65. In ———— form of Vertical Marketing System, one member of the channel is large and powerful enough to coordinate the activities of the other members without an ownership stake.

a) Corporate VMS

b) Administered VMS

c) Contractual VMS

d) None of the above

66. In ———— form of Vertical Marketing System, independent firms are joined together by contract for their mutual benefit.

a) Corporate VMS b) Administered VMS

c) Contractual VMS d) None of the above

67. ________ is a form of contractual VMS.

a) Retailer cooperative b) Franchise organization

c) Both a & b d) None of the above

68. In ———— type of contractual VMS, a group of retailers buy from a jointly owned wholesaler.

a) Retailer cooperative b) Franchise organization

c) Both a & b d) None of the above

69. In ———— type of contractual VMS, a producer licenses a wholesaler to distribute its products.

a) Retailer cooperative **b) Franchise organization**

c) Both a & b d) None of the above

70. Hybrid marketing channel network is one of the most important and widely used systems among types of ————

a) Vertical marketing channels

b) Horizontal marketing channels

c) E- marketing channels

d) Corporate marketing channels

71. ———— means that a company or site offers to transact or facilitate the selling of product and service online.

a) E-purchasing **b) E-commerce**

c) Corporate marketing d) All the above

72. ———— means companies decide to purchase goods services and information from various online suppliers.

a) E-purchasing b) E-commerce

c) Corporate marketing d) All the above

73. ———— is the method used to assess the degree of integration.

a) Price correlations

b) Spatial price differential and Transportation costs

c) Ravallion method

d) All the above

74. Which of the following is a method to assess market integration?

a) Co-integration approach b) Partially Bound models

c) Both a & b d) None of the above

75. According to ___________ ratio of total value of goods marketed to the marketing cost may be used as a measure of marketing efficiency.

a) Shepherds approach b) Acharya method

c) Both a & b d) None of the above

76. ___________ is an ideal measure of marketing efficiency comparing the marketing cost, marketing margin, price received by the farmer and consumer price.

a) Shepherds approach **b) Acharya method**

c) Both a & b d) None of the above

77. An efficient marketing system implies that price spread or marketing margin is fairly __________

a) High

b) Less

c) Maximum

d) Efficiency and price spread are not related

8

Supply Chain of Agricultural Marketing

1. **Principles of supply chain management are**

    a) Flexibility b) Reliability

    c) Visibility d) Innovation

2. **Supply Chain Management aims at reducing**

    a) Marketing Cost b) Role of intermediaries

    c) Lead time **d) Both A and C**

3. **A set of linked activities that works to improve a product while linking commodity producers to processors and markets.**

    a) Supply chain **b) Value chain**

    c) Marketing channel d) Marketing functions

4. **Supply chain stage refers to the steps taken**

    a) To deliver food from the farmer to the consumer

    b) To deliver food from its input supplier to the farmer

    c) To deliver food from its input supplier to the consumer

    d) To deliver food from the farmer to the distribution

5. **The supply chain includes the activities of the producer with connection to their suppliers**

    a) Downstream part b) Internal part

    c) External part **d) Upstream part**

6. **The internal part of the supply chain includes all of the**

    a) In-house processes used in transforming the inputs into the organization's outputs

    b) Activities of the producer with connection to their suppliers

    c) Activities of the producer with connection to the customers

    d) Activities of the input supplier with connection to the customers

7. **Supply chain management contains..........**
   a) Material handling b) Transportation
   c) Storage **d) All the above**
8. **The supply chain concept originated in the discipline of**
   a) Marketing b) Operations
   c) Logistics d) Production
9. **The downstream part of the supply chain includes _______________**
   a) Activities of the producer with connection to their suppliers
   b) Activities of the producer with connection to the customers
   c) Activities in transforming the inputs into the organization's outputs
   d) Activities of the input supplier with connection to the customers
10. **Sub-scheme of Integrated Scheme for Agricultural Marketing (ISAM) for holistic development of agricultural value chain**
    a) DMI **b) AMI**
    c) FMI d) IAM
11. **___________ plays important role in supply chain management**
    a) Marketing b) Finance
    c) Information system d) None of these
12. **Picking up of the product is a part of_______________when a customer return the product if it is not matching with ordered one**
    a) Sales management b) Retail Management
    c) Supply chain management d) Delivery Management
13. **In the present competitive world, every organisation want give proper service to the customers with the help of.......**
    a) Supply chain management b) Sales management
    c) Retail Management d) None of these
14. **Handling the material within the production process in the organisation is a part of ___________**
    a) Production b) Finance
    c) Supply chain management d) Marketing
15. **A set of linked activities that works to improve a product while linking commodity producers to processors and markets.**
    a) Supply chain **b) Value chain**
    c) Marketing channel d) Marketing functions

15. **Upgraded information about the flow of products, markets and technologies is the benefit of ______________**

    a) Value Chain Management   b) Supply Management

    c) Supply chain Management   d) Market Information

16. **The term "Value Chain" is defined by ______________**

    a) Samuelson   b) Bertrand

    c) Michael Porter   d) Robinson

17. **Supply chain is**

    a) Broader than marketing channels

    b) Both are same

    c) Lesser than marketing channels

    d) Both are different

19. **Pre and post inventory can be kept in the control with the help of...**

    a) Production   **b) Supply chain management**

    c) Marketing   d) Finance

20. **Example for money market**

    a) Shares   **b) Treasury bonds**

    c) Debentures   d) Mutual fund investments

21. **Examples for capital market**

    a) Inter-corporate deposits   b) Treasury bonds

    c) Debentures   d) Commercial papers and bills

22. **Financial instruments based on ownership securities consist of**

    a) Options   b) Derivatives

    c) Preference shares   d) Forward contracts

23. **One of the following is not the debt securities based financial instruments**

    a) Non-convertible debentures   b) Convertible debentures

    c) Equity share   d) Deep discount bonds

24. ________ is not a sources of financial information

a) Kothari's Economic and Industrial Guide of India

b) Stock Exchange Directory

c) Times of India Directory

d) Integrated Government Online Directory

25. NCDEX stands for

a) National Commodity & Derivatives Exchange Limited

b) National Common Derivatives Exchange

c) National Commodity Exchange

d) National Common Debentures Exchange

26. ___________ is the solution to all farming-related problems

a) NSEL Mandi b) XIMC Mandi

c) UCX Mandi **d) NCDEX Mandi**

27. The agency started with the responsibility of buffer stocking is

a. Commission for Agricultural Costs and Prices

b. Food Corporation of India

c. National Cooperative Union of India

d. Central Warehousing Corporation of India

28. _____________ exist when a company distributes its commodities or products through a channel that contains one or more resellers.

a) Indirect marketing b) Direct marketing

c) Multi-level marketing d) Integrated marketing

29. The abnormal increase in buffer stock of FCI in recent times is due to

a) Procurement prices are above market prices

b) Additional storage facility created by FCI

c) Transport bottlenecks in wheat producing regions

d) Series of bumper harvest

30. Sub-scheme of Integrated Scheme for Agricultural Marketing (ISAM) for holistic development of agricultural value chain

b) DMI **c) AMI**

d) FMI e) IAM

31. Asian largest supply chain management is ___________

a) ITC **b) FCI**

c) APEDA d) FAO

32. Many time people has confusion between SCM and________

a) Finance **b) Transportation**

c) Marketing d) Human resource

33. In a SC, Material flows in one direction while ________ from in both direction

a) Process **b) Information**

c) Product d) Semi-finished Goods

9

Market Integration Horizontal, Vertical and Conglomeration

1. **The expansion of firms by consolidating supplementary marketing functions and activities under a single management**

   a) Market Segment b) Market integration

   c) Market channel d) Market Regulation

2. **When a firm gains control over other firms which performing similar marketing functions are __________**

   a) Vertical co-ordination b) **Horizontal integration**

   c) Vertical integration d) conglomeration

3. **When a firm performs more than one activity in the order or sequence of the marketing process is known as _________**

   a) Vertical co-ordination b) Conglomeration

   c) Horizontal integration **d) Vertical integration**

4. **A combination of agencies or activities not directly related to each other, may when it operates under a united management**

   a) Conglomeration b) Vertical co-ordination

   c) Vertical integration d) Horizontal integration

5. **Help companies expand in size, diversify product offerings, reduce competition, and expand into new markets.**

   a) Supply chain management **b) Horizontal integration**

   c) Co-operatives d) Market structure

6. **Helps in boosting profit and allow companies more immediate access to consumers.**

   a) Vertical integrations b) Supply chain management

   c) Retail markets d) Co-operatives

7. When a juice company acquires a mango orchard

a) Backward Integration b) Amalgamation

c) Conglomeration d) Merger

8. Forward Integration occurs when

a) Restaurant owner buys a poultry farm

b) Wheat flour manufacturing company decides to take control of the post-production process

c) Poultry farmer opens a restaurant

d) Wheat flour manufacturing company decides to take control of the maida production company

9. Balanced integration

a) Backward and forward integration

b) Horizontal and vertical integration

c) Conglomeration

d) Forward and conglomeration

10. The degree of price transmission between either vertically or spatially related markets, as a proxy for marketing efficiency.

a) Market structure **b) Market integration**

c) Marketing functions d) Price spread

11. Shorter marketing channel would benefit

a) Producer b) Consumer

c) Both A and B d) Traders

12. The synchronization of successive stages of production and marketing, with respect to quantity, quality, and timing of product flows.

a) **Vertical coordination** b) Horizontal integration

c) Vertical integration d) Horizontal coordination

13. Extent of market integration influences

a) Market conduct b) Market efficiency

c) Both d) None of the above

14. Any firm or agency gain control of other firms or agencies which performing similar marketing functions at the same level in marketing sequence

a) Vertical integration **b) Horizontal integration**

c) Both d) Conglomeration

15. Horizontal integration advantageous

a) Consumers/buyers **b) Member of the group**

c) Producers d) None of the above

16. The degree to which a firm owns its upstream suppliers and its downstream buyers

a) Vertical integration b) Horizontal integration

c) Both d) Conglomeration

17. Wholesaler assuming function of retailing is a

a) Forward vertical integration

b) Vertical integration

c) Backward integration

d) Conglomeration

18. Processing firm undertake function of assembling /purchasing produce from village is a

a) Forward vertical integration b) Vertical integration

c) Backward integration d) Horizontal integration

19. An automobile company is an example of

a) Forward vertical integration b) Vertical integration

c) Backward integration d) Horizontal integration

20. Modern retail stores exhibits

a) Forward vertical integration b) Vertical integration

c) Horizontal integration d) Both b and c

21. A oil mill opening its retail outlet is

a) Horizontal integration **b) Forward integration**

c) Backward integration d) Conglomeration

22. Kisan company engaging in tomato procurement directly from farmers is

a) Forward integration b) Conglomeration

c) Vertical integration d) Backward integration

23. Which of the following is an example of horizontal integration?

a) A wholesale function assuming the function of retailing

b) Ramoji group of industries

c) Activities performed by Britannia company

d) Cooperative farming

24. ______________ is an example of vertical integration?

a) A wholesale function assuming the function of retailing

b) Ramoji group of industries

c) Cooperative farming

d) All the above

25. Which of the following is an example of forward integration?

a) A wholesale function assuming the function of retailing

b) Ramoji group of industries

c) Cooperative farming

d) All the above

26. ______________ is an example of backward integration?

a) A wholesale function assuming the function of assembling the produce

b) Ramoji group of industries

c) Cooperative farming

d) All the above

27. ______________ is an example of conglomeration?

a) A wholesale function assuming the function of assembling the produce

b) Ramoji group of industries

c) Cooperative farming

d) All the above

28. Services provided by marketing channels to their customer in a collective way is classified as

a) Functional integration
b) Product integration
c) Channel integration
d) Location integration

29. Horizontal integration is the process of a company increasing production of goods or services at the same part of the supply chain

a) Forward integration
b) Horizontal integration
c) Conglomeration
d) Vertical integration

30. Horizontal integration occurs when

a) Firm creates singular country production facilities, each of which produces different good or goods
b) Firm creates multiple production facilities, each of which produces the same good or goods.
c) Firm creates multiple production facilities, each of which produces different good or goods
d) Firm creates singular country facilities, each of which produces the same good or goods

31. The extent of _______________in a market may be assessed by counting the number of functions performed by each firm in the market

a) Forward integration
b) Vertical integration
c) Conglomeration
d) Backward integration

32. The extent of _________________may be measured by studying the number of firms performing the same marketing function but operating under one common management

a) Horizontal integration
b) Horizontal integration
c) Conglomeration
d) Backward integration

33. The measurement or assessment of market integration may be attempted at ——————————

a) Integration among firms of a market
b) Integration among spatially separated markets
c) Both
d) Integration among commodities in the market

34. **_________________ occurs when all the assets and decisions of a firm are completely assumed by another firm**

a) **Ownership integration** b) Contract integration

c) Market integration d) Backward integration

35. **___________________ involves an agreement between two firms while taking certain decisions each firm retains its own separate identity**

a) Ownership integration **b) Contract integration**

c) Market integration d) Backward integration

36. **If a processing firm buys a wholesale firm, it is an example of ——————————**

a) Ownership integration b) Contract integration

c) Market integration d) Backward integration

37. **Tie up of a dhal mill with pulse traders for supply of pulse grains is an example of ————**

a) Ownership integration **b) Contract integration**

c) Market integration d) Backward integration

38. **Risk is reduced through diversification exists in**

a) Forward integration b) Vertical integration

c) Conglomeration d) Backward integration

39. **Effects of vertical integration is/are**

a) Reduced risk through improved market coordination

b) More profits by additional functions and low costs through operational efficiency

c) Improvement in bargaining power and influencing prices

d) All the above

40. **NAFED is an example of**

a) Forward integration b) Vertical integration

c) Conglomeration d) Backward integration

41. **Disadvantages of the Horizontal integration**

a) Destroyed value b) Legal repercussions

c) Reduced flexibility **d) All the above**

42. Another name of horizontal integration

a) Parallel integration b) Similar integration

c) Competitor integration **d) Lateral integration**

43. Another name of horizontal integration

a) Parallel integration b) Similar integration

c) Competitor integration **d) Lateral integration**

44. When a firm grows by merging with another firm at the same stage of production and in the same line of activity it is often called

a) Forward integration b) Vertical integration

c) Horizontal integration d) Backward integration

45. Which of the following may be reasons for horizontal integration as a method of growth?

a) Need for suppliers to produce better quality components for the manufactured product

b) Need for manufacturer to have more retail display of products on high streets

c) Significant technical economies of scale associated with plant size

d) Need to reduce risk by operating in different product markets

46. Which of the following is an example of a firm growing by forward vertical integration?

a) A knitwear business acquires another knitwear business

b) A fresh vegetable merchant acquires a restaurant

c) Shoe manufacturer acquires a leather works

d) A steakhouse restaurant acquires a cattle-breeding business

47. Companies that seek to strengthen their positions in the market and enhance their production or distribution stage use ____________

a) Forward integration b) Vertical integration

c) Horizontal integration d) Backward integration

48. Involving a firm in a totally unrelated business

a) Forward vertical integration b) Lateral integration

c) Conglomerate integration d) Backward vertical integration

49. Towards the raw material supplier

a) Forward vertical integration b) Lateral integration

c) Conglomerate integration **d) Backward vertical integration**

50. Towards the final consumer

a) Forward vertical integration

b) Lateral integration

c) Conglomerate integration

d) Backward vertical integration

51. Involves firms in different product areas, but with some common elements

a) Organic growth **b) Lateral integration**

c) Conglomerate integration d) Backward vertical integration

10

Agricultural Price Policy

1. **In developed countries, the major objective of price policy is to____________**
   a) **Prevent drastic fall in agricultural income**
   b) Increase the agricultural production
   c) Both a & b
   d) None of the above
2. **In developing economies it is to ————**
   a) Prevent drastic fall in agricultural income
   b) **Increase the agricultural production**
   c) Both a & b
   d) None of the above
3. **The important objectives of the new agricultural policy are ———**
   a) Augment facilities for All-Round Development
   b) Infrastructural Development
   c) Reviving and strengthening Co-operatives and local communities
   d) **All the above**
4. **The important measures or features of new agricultural policy are ————**
   a) Raise the rate of capital formation in agricultural sector
   b) Raise the volume of public investment
   c) Enhance the flow of credit to the agricultural sector
   d) **All the above**
5. **In ———— the terms of trade were purposively kept unfavourable for the agricultural sector.**
   a) **Negative price policy** b) Positive price policy
   c) Both a & b d) None of the above

6. **———— consists of light taxes on the agricultural sector and assure the farmer of a fair price for his produce.**

   a) Negative price policy **b) Positive price policy**

   c) Both a & b d) None of the above

7. **The major shortcomings of the agricultural price are ————**

   a) Inadequate coverage of procurement facility

   b) Failure of remunerative price and/or subsidized inputs with increasing rate of costs

   c) Ineffective Public Distribution System

   d) All the above

8. **Price volatility leads to ————**

   a) Uncertainty and risks

   b) Weaken the agricultural performance and have

   c) Negative impact on income and welfare of the farmers and the rural poor

   d) All the above

9. **Futures markets can help only if**

   a) The government interventions do not hinder the normal, efficient and competitive flow of commodities in the economy

   b) A sufficiently large part of physical trade should be handed over to the private sector and prices should be allowed to clear the market

   c) Both a & b

   d) None of the above

10. **Commodities exchanges would need to upgrade their ————**

    a) Rules and regulations of trading procedures

    b) Delivery system, trade supervision

    c) Capacity to design and introduce futures contracts

    d) All the above

11. **The most popular exchanges in the World are ————**

    a) CME Group b) Kansas City Board of Trade

    c) London Metal Exchange **d) All the above**

12. The categories of trading commodities include ————

a) Energy, Metals, Livestock & Meat, Agricultural commodities

b) Metals, Livestock and Meat, Agricultural commodities

c) Energy, Metals, Livestock and Meat

d) Energy, Metals, Agricultural commodities

13. ———— was established by Government of India to co-ordinate the agricultural marketing of various agencies and to advise the Central and State Governments on the problems of agricultural marketing.

a) Directorate of marketing and inspection

b) Agricultural Produce Market Committee

c) Food Corporation of India

d) None

14. ———— agency in the public sector engaged in building large scale storage/ warehousing capacity

a) Food Corporation of India (FCI)

b) Central Warehousing Corporation (CWC)

c) State Warehousing Corporation (SWC)

d) All the above

15. Central Warehousing Corporation was set up in ————

a) 1962 b) 1966

c) 1952 d) 1958

16. Which of the following is/are the advantage/s of co-operative marketing?

a) Increases bargaining power of the farmers

b) Provision of credit

c) Storage facilities

d) All the above

17. ———— is the scheme implemented by Directorate of Marketing and Inspection

a) Market research & planning b) Market planning and design

c) Both a & b d) None

18. Regulated markets are managed by ————

a) **Agricultural Produce Market Committee**

b) Directorate of Marketing and Inspection

c) Food Corporation of India

d) Central Warehousing Corporation

19. The economic cost of the process of food procurement and distribution includes ———— components.

a) Price paid to the farmers b) Procurement operations

c) Cost of distribution **d) All the above**

20. The difference between economic cost of food grains and the issue price of FCI is equivalent to ————

a) Food subsidy b) Farmers' price

c) Market price d) None

21. To accelerate economic growth, a ———— agricultural price policy has been practiced by a large number of countries in the early stages of their development.

a) Negative price policy b) Positive price policy

c) Both a & b d) None of the above

22. ———— policy is considered necessary in the context of the realisation that unless the agricultural sector attains some critical minimum rate of growth, it would not be possible to attain the general targets of economic growth and development.

a) Negative price policy **b) Positive price policy**

c) Both a & b d) None of the above

23. ____________ is an effect of Agricultural Price Policy.

a) Incentive to Increase Production

b) Increase in the level of income of Farmers

c) Price Stability

d) All the above

24. Agricultural Price Policy benefited the agro industries by ————

a) Stabilizing the prices of agricultural commodities

b) Provision for adequate quantity of raw material at reasonable price

c) Both a & b

d) None of the above

25. ———— is the important policy change occurred during 1990s that affected incentive structure of the agriculture sector.

a) Domestic economic reforms b) WTO Agreement on Agriculture

c) Both a & b d) None of the above

26. The economic reforms that reduced the levels of industrial protection have ———— the incentives to agriculture.

a) Improved b) Degraded

c) Worsen d) Not affected

27. The State Agricultural Produce Marketing (Development and Regulation) Act was enacted in the year ————

a) 2005 **b) 2003**

c) 2001 d) 2006

28. The Food Safety and Standards Act was enacted in the year ————

a) 2005 b) 2003

c) 2001 **d) 2006**

29. The Forward Markets (Regulation) Act was enacted in the year ————

a) 1952 b) 1963

c) 1958 d) 1960

30. ———— performs the functions of advisory, monitoring, supervision and regulation in futures and forward trading.

a) Securities and Exchange Board

b) National Commodity & Derivatives Exchange Limited

c) Forward Markets Commission

d) None of the above

31. The Food Safety and Standards Act consolidates ————

i) The Prevention of Food Adulteration Act, 1954;

ii) The Fruits Products Order, 1955;

iii) The Meat Food Products Order, 1973;

iv) The Vegetable Oil Products (Control) Order, 1947;

a) (i) and (iv) b) (i) (ii) and (iv)

c) (i) (ii) (iii) and (iv) d) (i) only

32. E-NAM was established in the year ——————

a) 2016 b) 2018

c) 2014 d) 2012

33. The minimum support price is__________

a) The price at which the consumers can buy crops from farmers

b) The price at which the Government buys grains from the farmers

c) The price at which farmers sell their crops to the consumers for assured profit

d) The price at which the crops can be purchased by the processors

34. When was the concept of MSP introduced in India?

a) 1996 **b) 1966**

c) 1969 d) 1986

35. The concept of MSP introduced for _________ in India?

a) Cotton b) Rice

c) Sugarcane **d) Wheat**

36. Who fixes the Minimum Support Price for the crops?

a) Commission for Agricultural Costs and Prices

b) Government of India

c) Cabinet Committee on Economic Affairs

d) NITI Ayog

37. E-NAM is an India based electronic trading portal which networking the existing _________ to create a unified national market for agricultural commodities.

a) Wholesale markets b) Cooperative markets

c) APMC markets d) Farmers markets

38. Regulated markets aim at the development of the marketing structure to have ——————

a) Ensure remunerative price to the producer of agricultural commodities,

b) Reduce non-functional margins of the traders and commission agents, and

c) Narrow down the price spread between the producer and the consumer

d) All the above

39. The Central Warehousing Corporation was set up in ————

a) 1950 **b) 1957**

c) 1962 d) 1964

40. Which of the following is true statement?

i) The Agricultural price policy aims at providing assured price to the farmers.

ii) It aims at inducing farmers to bring in their surplus produce to the market.

a) i only b) ii only

c) Both d) None

41. What is MSP for rice for the year 2021-2022?

a) 1960 b) 1980

c) 1940 d) 1950

42 ———— was established by the Government of India to coordinate the agricultural marketing of various agencies and to advise the Central and State Governments on the problems of agricultural marketing.

a) Directorate of Marketing and Inspection

b) Food Corporation of India

c) Agriculture Produce Market Committee

d) None of the above

43. The activity of the Directorate of Marketing and Inspection is ————

a) Promotion of grading and standardization of agricultural and allied commodities

b) Statutory regulation of markets and market practices

c) Market extension

d) All the above

44. _______ is the scheme implemented by the Directorate of Marketing and Inspection.

a) Market research and planning b) Market planning and design

c) Both a & b d) None of the above

45. ———— intervene in the market to make direct purchase from the farmers at the support prices.

a) Food Corporation of India

b) Central Warehousing Corporation

c) State Warehousing Corporation

d) All the above

46. ———— is the main agency in the public sector engaged in building large scale storage/warehousing capacity.

a) Food Corporation of India

b) Central Warehousing Corporation

c) State Warehousing Corporation

d) All the above

47. Price of rice sold to the consumers through PDS is:

a) Ceiling Price b) Levy price

c) Minimum Support Price **d) Retail Issue price**

48. Food grains prices supplied through fair price shops and rationing at subsidized rates are issue prices.

a) Minimum Support price **b) Issue price**

c) Procurement Price d) Statutory minimum price

49. Price is below open market prices and always higher than procurement prices.

a) Ceiling Price b) Procurement price

c) Issue price d) Minimum Support Price

50. Regulated market ensures

a) Minimum Support price b) Procurement price

c) Issue Price **d) Fair price**

51. ____________ refers to the price at which government procures from producers to maintain buffer stocks

a) Ceiling price b) Minimum Support price

c) Procurement price d) Issue Price

52. Which of the following is true statement?

a) Procurement price < MSP

b) Issue Price > Market price

c) Issue price < Procurement Price

d) Issue price < MSP

11

Risks in Marketing and Forward and Spot Market

1. **In ___________ market, future buying and selling of commodities will take place at current time**

   a) Forward b) International

   c) Perfect d) Spot

2. **Forward Contracts (Regulation) Act**

   a) 1952 b) 1962

   c) 1958 d) 1963

3. **____________ had also recommended strengthening of existing commodity exchanges and forward markets commission**

   a) Kuhsro committee b) Rangarajan committee

   c) Forward market committee **d) Kabra committee**

4. **The functions of Forward Markets Commission are**

   a) Recognition of associations conducting forward trading

   b) Keep forward markets under observation

   c) Submit periodical reports to Government

   d) All the above

5. **Future Trading in India was started at**

   a) 1875 b) 1785

   c) 1857 d) 1757

6. **A measure of frequency and severity of price movement in a given market**

   a) Price discovery **b) Price volatility**

   c) Price determination d) Price elasticity

7. **The remedial measures for eliminating the marketing problems are ________________**

   a) Government legislations

   b) Organisation of cooperative marketing,

   c) Reduction and regulation of market charges

   d) All the above

8. **Forward market is otherwise called as**

   a) Future market b) Regulated market

   c) Closed market d) Black Market

9. **Black market is otherwise called as**

   a) Future market b) Regulated market

   c) Closed market d) Future market

10. **The forward market is especially well-suited to offer hedging protection against**

    a) Translation risk exposure **b) Transactions risk exposure**

    c) Political risk exposure d) Taxation

11. **A black market for currencies exists in some countries where.**

    a) There is a huge balance of payments surplus

    b) There is a huge trade surplus

    c) Currencies are pegged and exchange controls exist

    d) All of the above

12. **__________ is not a characteristic of speculation.**

    a) Profit motive b) Exchange rate fluctuation us

    c) Hedging d) Risk taking

13. **Speculation in foreign exchange markets entails.**

    a) Covering in the forward market

    b) Covering in the money market

    c) Hedging in the option market

    d) Buying in the current spot market and selling in the future spot market

14. A forward rate is equal to a future spot rate if foreign exchange markets are

a) Controlled by the government

b) **Efficient**

c) Controlled by speculators

d) Partially controlled by the International Monetary Fund

15. If the spot rate of the Deutsche mark is $30 and the six month forward rate of the mark is $32, what is the forward premium or discount on an annual basis?

a) Premium; about 14.5% b) Discount; about 14.5%

c) Premium; about 13.3% d) Discount; about 13.3%.

16. If the spot rate of the Deutsche mark is $.32 and the six month forward rate is $.30, what is the forward premium or discount on an annual basis?

a) Discount; 11.5% b) Premium; 11.5%

c) Premium; 12.5% d) Discount; 12.0%

17. The basis is defined as:

a) Spot price minus forward price

b) Futures price minus forward price

c) Futures price minus spot price

d) Spot price minus futures price

18. The basis must equal __________ at the delivery date for the futures contract.

a) Zero b) The spot price of the contract

c) A positive value d) A negative value

19. A futures contract is

a) An agreement to buy or sell a specified quantity of a particular asset during a given period at the spot price

b) A marketable obligation to buy or sell a specified quantity of a particular asset during a given period for a given price

c) An option to buy or sell a specified quantity of a particular asset during a given period for a given price

d) A marketable obligation to buy or sell a specified quantity of a particular asset during a given period at a price to be determined in the future

20. If a farmer buys a corn option on futures

a) Farmer must deliver the corn at a fixed price

b) Farmer must deliver the corn at the market price

c) Farmer has the right to deliver the corn, and will do so only if the price is favorable

d) Farmer must accept delivery of the corn at a fixed price

21. The forward market is especially well-suited to offer hedging protection against

a) Transactions risk exposure b) Political risk exposure

c) Taxation d) Translation risk exposure

22. Assume that a Big Mac hamburger is selling for $ 1.99 in the United Kingdom, the same hamburger is selling for $2.71 in the United States, and the actual exchange rate (to buy $1.00 with British pounds) is 0.63. According to ———————— the British pound is————— US dollar

a) Purchasing-power parity; undervalued

b) Interest-rate parity; undervalued

c) Purchasing-power parity; overvalued

d) Interest-rate parity; overvalued

23. What is the name of the price pattern, if futures prices are increasing in time to maturity?

a) Backwardation

b) Fair value

c) Contango

d) Futures are only traded for specific maturity date, so that it is impossible to draw a curve of futures prices against time to maturity

24. Consider futures contracts with the S&P 500 index as the underlying. How are these contracts settled at the maturity date?

a) The short position in the futures contract has to make sure to deliver all 500 stocks that are represented in the S&P 500 index in the appropriate numbers to exactly replicate the index composition

b) The short position in the futures contract has to deliver shares in an index fund that tracks the S&P 500

c) The long position in the futures contract has to deliver shares in an index fund that tracks the S&P 500

d) S&P 500 futures are cash settled, that is only money changes hands according to the difference between the price specified in the futures contract and the current value of the S&P 500

25. On a future exchange, one future contract is traded where bot the long and short parties in the open interest is

a) No exchange

b) Decrease by two

c) Decrease by one

d) Increase by one

26. Which of the following is true?

a) The optimal hedge ratio is the slope of the best fit line when the spot price (on y axis) is regressed against the futures price (on x axis)

b) The optimal hedge ratio is the slope of the best fit line when the future price (on y axis) is regressed against the spot price (on x axis)

c) The optimal hedge ratio is the slope of the best fit line when the change in the spot price (on y axis) is regressed against the change in the futures price (on x axis)

d) The optimal hedge ratio is the slope of the best fit line when the change in the future price (on y axis) is regressed against the change in the spot price (on x axis)

27. The basis at a given instant of time is defined as spot price minus future price at that instant. If the basis strengthens unexpectedly, which of the following is true?

a) The shorter hedger's position improves

b) The shorter hedger's position worsens

c) The shorter hedger's position sometime improves and sometime worsens

d) The shorter hedger's position stays the same

28. _______________ is not generally considered a benefit of hedging?

a) It reduces one or more aspects of business risk.

b) It allows prices to be locked in advance.

c) The costs of hedging are paid by the speculators.

d) It can stabilize profits.

29. Which of the following is not correct concerning futures contracts?

a) Entails an obligation rather than an option.

b) Contract price is set at the beginning of the contract.

c) Contracts are exchange-traded.

d) Gains or losses are recorded at contract expiration.

30. What happens to the price of a futures contract as expiration draws closer?

a) It exceeds the spot price of the asset.

b) It is exceeded by the spot price of the asset.

c) It approaches the spot price of the asset.

d) There is no relationship between futures price and spot price as the contract approaches expiration.

31. The process of marking a futures contract to market means that:

a) The profitability of the contract is locked in from the onset of the contract.

b) The amount of commodity to be delivered changes as prices change.

c) Contracts are closed out as soon as they become unprofitable.

d) Profits or losses are posted to the contract daily.

32. The basic difference between speculators and hedgers in futures contracts is that speculators:

a) Will profit regardless of the direction of price change.

b) Are not protecting their commodity holdings.

c) Are concerned only with long-term price movements.

d) Take a position in more than one commodity at a time.

33. When two borrowers engage in a currency swap, they agree to:

a) Trade one currency for another, thus avoiding the foreign exchange market.

b) Make payments on each other's borrowings in a different currency.

c) Pay to each other any depreciation or appreciation of the currency.

d) Exchange fixed-rate interest payments for variable-rate interest payments.

34. If most hedging acts to reduce risk, managers should expect that hedging will:

a) Increase profits.

b) Decrease profits.

c) Increase the firm's stock price.

d) Stabilize the firm's dividend payout.

35. Why are most futures contracts not settled through delivery of the product?

a) Most contracts are settled through the margin account.

b) Most contracts expire with neither party having an obligation to the other party.

c) Most participants cancel their futures contracts through purchase of an option contract.

d) It is easier and cheaper to settle in cash or by offset.

36. Which of the following is true regarding firm's hedging for risk reduction?

i) Fluctuations in commodity prices, interest rates, or exchange rates can make planning difficult and can throw companies badly off course.

ii) Financial managers look for opportunities to manage these risks, and a number of specialized instruments have been invented to help them.

iii) The specialized instruments which were invented to help financial managers manage risks are collectively known as derivative instruments.

a) i　　b) ii

c) iii　　**d) i, ii and iii.**

37. How can companies use swaps to change the risk of securities that they have issued?

i) Swaps allow firms to exchange one series of future payments for another.

ii) The firm might agree to make a series of regular payments in one currency in return for receiving a series of payments in another currency.

a) i　　b) ii

c) Both i and ii are correct.　　d) Neither i nor ii is correct.

38. Speculation and hedging are important ways to minimize

a) Production risk b) Transport problems

c) Price risk d) Technological risk

39. Manipulative practices to create conditions of artificial scarcity in the product market and lead to price rise

a) Illegitimate Speculation b) Arbitrage

c) Hedging d) Swap

40. Opposite sales or purchases in the futures market to offset the purchases or sales of physical products made in the cash market

a) Speculation **b) Hedging**

c) Arbitrage d) Swap

41. The purchases and sales in the cash and futures markets are made to protect against excessive price fluctuations

a) Speculation b) Arbitrage

c) Swap **d) Hedging**

42. Purchases and sales in the cash as well as in future markets are made with the objective of making profit.

a) Arbitrage **b) Speculation**

c) Swap d) Hedging

43. Financial derivatives include

a) Stocks b) Bonds

c) Futures d) None of the above

44. Which of the following is not a financial derivative?

a) Stock b) Futures

c) Options d) Forward contracts

45. Elimination of riskless profit opportunities in the futures market is

a) Hedging **b) Arbitrage**

c) Speculation d) Underwriting

e) Diversification

46. Options on individual stocks are referred to as

a) Stock options. b) Futures options.

c) American options. d) Individual options.

47. An instrument for managing the risk of interest rate that requires exchange of payment streams is a

a) Hedge. b) Forward contract.

c) Swap. d) Futures contract

48. ____________ is the exchange of a set of payments in one currency for a set of payments in another currency

a) Interest rate swap. **b) Currency swap.**

c) Swaptions. d) National swap.

49. __________ involves the exchanging one set of interest payments for another set of interest payments

a) Currency swap. **b) Interest rate swap**

c) Swaptions. d) National swap.

50. Advantage of using swaps to eliminate the risk of interest rate is that swaps ___________

a) Have better accounting treatment than options.

b) Are less costly than rearranging balance sheets.

c) Are less costly than futures.

d) Are more liquid than futures

51. Disadvantage of swaps is that they ___________

a) Lack liquidity

b) Are difficult to arrange for a counterparty

c) Suffer from default risk

d) All of the above

52. A financial contract that obligates one party to exchange a set of payments it owns for another set of payments owned by another party is called a

a) Hedge b) Call option

c) Put option **d) Swap**

53. A bipartite agreement between farmer and processing and/or marketing firms for the production and supply of the agricultural products quantity at a specified price and time.

a) Contract farming b) Collective farming

c) Co-operative farming d) Co-ordinated farming

54. Production supporting mechanism by an agreement between farmer and/or marketing firms at predetermined prices.

a) Co-operative farming **b) Contract farming**

c) Subsistence farming d) Mixed farming

55. A centralized processor purchase produce from a large number of small farmers is __________ farming

a) Intermediary contract farming b) Buyer contract farming

c) Informal contract farming **d) Centralized contract farming**

56. In ____________ farming where the sponsor may lose the control of production and quality as well as prices received by farmers.

a) Centralized contract farming

b) Informal contract farming

c) Intermediary contract farming

d) Buyer contract farming

57. __________ are agreements of contracts to purchase or sell a specified amount of a given commodity at a fixed price in a specified future date

a) Contract farming **b) Commodity futures**

c) Agreement farming d) Intermediary contract farming

58. The intensity of the contractual arrangement varies according to the depth and complexity of the following

a) Market provision: b) Resource provision

c) Management specifications **d) Social provision**

59. An agreement of purchase between Cotton farmer and ginning industry is an example of __________

a) Bilateral Contract Farming b) Mixed farming

c) Trilateral Contract Farming d) Forwards

60. A type of farming with a risk of default by both the farmer and the promoter is______

a) Contract farming b) Mixed farming

c) Share Farming d) Co-operative farming

61. Which state has launched contract farming scheme to energise agriculture?

a) Punjab **b) Haryana**

c) Uttar Pradesh d) West Bengal

62. In _______________ farming, a centralized processor and/or packer buys from a large number of small farmers

a) Intermediary contract farming b) Buyer contract farming

c) Informal contract farming **d) Centralized contract farming**

63. The ___________ model consists of individual entrepreneurs or small companies.

a) Informal contract farming b) Intermediary contract farming

c) Buyer contract farming d) Centralized contract farming

64. When the owner of a product gives legal license to the business man in particular territories to make profit for the owner, it is termed as___________

a) Joint Ownership b) Contract Farming

c) Conglomeration **d) Franchising**

12

Marketing Institutions

1. **Government is concerned with the marketing system for**

   a) The quality of commodities at the lowest price

   b) Maximum price for surplus produce

   c) Steady income from marketing of commodities

   d) **Safeguard the interests of the groups associated in marketing**

2. **Which of the following statements are correct ?**

   i) The Jute Corporation of India is providing minimum price support that assists jute cultivators in all the states of India.

   ii) The company purchases jute at the fixed minimum price when market prices declined and later sells to jute mills.

   iii) The jute industry suffers from wide fluctuations in supply and price due to changes in the cultivated area and due to the effects of the factors associated with weather.

   a) i and ii only b) i and iii only

   c) ii and iii only d) All of the above

3. Leading international trading company of the Government of India is mainly for export and import operations

   a) Indian Trade Corporation

   b) State Trading Corporation of India Limited

   c) Foreign Trade Corporation

   d) International Trade Corporation

4. Co-existing of both private traders and government

   a) Partial State Trading b) Complete State trading

   c) Co-operative State trading d) Associated State trading

5. **Complete state trading of governmental agency performs**
   a) Necessitates the outlay of huge finance
   b) Creating an artificial scarcity by hoarding with a view to raising prices
   c) Management of human resource
   d) Provision of storage facilities
6. **Tobacco Board India is located at**
   a) Nellur **b) Guntur**
   c) Hyderabad d) Kurnool
7. **Central Silk Board (CSB) is located at**
   a) Varanasi b) Mysore
   c) Bangalore d) Bhoodan Pochampally
8. **Tea Board of India is located at**
   a) Kolkata b) Jorhat
   c) Meghalaya d) Dibrugarh
9. **National Oilseeds and Vegetable Oils Development Board is located at**
   a) Gurgaon b) Gwalior
   c) Goriwala d) Jalgaon
10. **Cotton corporation of India was established in**
    a) 1970 b) 1975
    c) 1978 d) 1972
11. **DMI is located at**
    a) Hyderabad b) Ahmedabad
    c) Secunderabad **d) Faridabad**
12. **Shorter marketing channel would benefit**
    a) Producer b) Consumer
    c) Both A and B d) Traders
13. **National Institute of Agricultural Marketing (NIAM) is a premier National level Institute is located at**
    a) Jodhpur **b) Jaipur**
    c) Udaipur d) Jaisalmer

14. Central Agmark Laboratory is located at

a) Pune **b) Nagpur**

c) Kanpur d) Patna

15. A High Power Committee on Agricultural Marketing was constituted in

a) 1996 **b) 1992**

c) 1982 d) 1986

16. Food Corporation of India was setup in the year

a) 1965 b) 1955

c) 1956 d) 1966

17. Sale of agricultural goods and products from the farm straight to the consumer without intervening distributors or retailers

a) Forward marketing **b) Direct marketing**

c) Regulated marketing d) Vertical integration

18. Example for Direct marketing

a) Apni Mandi Initiative in Punjab

b) Rythu Bazar of Andhra Pradesh

c) Uzhavar Sandhais of Tamil Nadu

d) All of the above

19. Union Minister for Agriculture and Farmer's Welfare launched which quality mark logo at New Delhi?

a) **National Dairy Development Board**

b) National Dairy Development Authority

c) National Dairy Development Scheme

d) National Dairy Development Programme

20. Central Silk Board (CSB) is located at

a) Varanasi b) Mysore

c) Bangalore d) Bhoodan Pochampally

21. Tea Board of India is located at

a) Kolkata b) Jorhat

c) Meghalaya d) Dibrugarh

22. Tobacco Board India is located at

a) Nellur **b) Guntur**

c) Hyderabad d) Kurnool

23. National Oilseeds and Vegetable Oils Development Board

a) Gurgaon b) Gwalior

c) Goriwala d) Jalgaon

24. __________ involved in teaching, research and consultancy services and to conduct training courses to augment the agricultural marketing

a) DMI b) **NIAM**

c) AMI d) High Power Committee

25. ____________________ is not the objective of Agricultural Marketing Infrastructure

a) Developing and upgrading Gramin Haats as GrAMs

b) **Assistance for renovation of storage infrastructure**

c) Promote innovative and latest technologies in post-harvest

d) Develop alternative & competitive marketing channels

26. The agency started with the responsibility of buffer stocking is

a) Commission for Agricultural Costs and Prices

b) Food Corporation of India

c) National Cooperative Union of India

d) Central Warehousing Corporation of India

27. Problems related to farming can be sorted out by

a) XIMC Mandi b) NSEL Mandi

c) NCDEX Mandi d) UCX Mandi

28. APMC is aimed to ____________

a) **Protect the interest of the farmers**

b) Protect the interest of private sector in agriculture

c) Provide better suggestive measures to enhanced utilisation of green fertiliser

d) Protect the interest of the buyers of agricultural produce

29. **Primary international trading company of the Government of India is for export and import is ___________**

    a) International Trade Corporation

    b) Indian Trade Corporation

    c) Foreign Trade Corporation

    d) **State Trading Corporation of India Limited**

30. **Promotion of grading and standardization of agricultural and allied products under the Agricultural Produce (Grading & Marketing) Act. 1937**

    a) DMI b) NIAM

    c) MANAGE d) AMI

31. **National Institute of Agricultural Marketing has started functioning at**

    a) **Jaipur** b) Udaipur

    c) Jamshedpur d) Lucknow

32. **Sale of agricultural goods and products from the farm straight to the consumer without intervening distributors or retailers**

    a) Forward marketing **b) Direct marketing**

    c) Regulated marketing d) Indirect marketing

33. **Example for Direct marketing**

    a) Apni Mandi Initiative in Punjab

    b) Rythu Bazar of Andhra Pradesh

    c) Uzhavar Sandhais of Tamil Nadu

    d) All of the above

35. **When a company supply its products through a channel that includes one or more resellers, it is known as ________.**

    a) Indirect marketing b) Direct marketing

    c) Multi-level marketing d) Integrated marketing

13

International Trade

1. **International trade contributes and increases the world ________**

   a) Population b) Inflation

   c) Economy d) Trade Barriers

2. **Free international trade maximizes world output through________.**

   a) Countries reducing various taxes imposed.

   b) **Countries specializing in production of goods they are best suited for.**

   c) Perfect competition between countries and other special regions

   d) The diluting the international business laws & conditions between countries.

3. **Domestic company limits it's operations to __________ political boundaries.**

   a) International b) **National**

   c) Transnational d) Global

4. **Trade between two or more than two countries is known as ________.**

   a) Internal Business b) External Trade

   c) **International Trade** d) Unilateral Trade

5. **_____ refers to the tax imposed on imports.**

   a) Imported Tax b) **Tariffs**

   c) Subsidies d) Import Quotas

6. **_____ means selling the products at a price less than on going price in the market.**

   a) Quota b) Tariff

   c) Subsidy d) **Dumping**

7. _______ is the oldest International Trade theory.

a) Country Similarity Theory

b) Theory of Absolute Cost advantage

c) Product Life Cycle Theory

d) **Mercantilism Theory**

8. A voluntary export restraint is the opposite form of _____.

a) **Import quotas** b) International tariffs

c) Subsidies d) Dumping

9. _________ is a group of countries agree to abolish all trade restrictionsand barriers.

a) Common market b) Economic Union

c) Custom Union d) **Free Trade Area**

10. The full form of WTO is __________

a) World Tariff Organization b) **World Trade Organization**

c) Western Trade Organization d) World Transport Organization

11. ________ was replaced by WTO on January 1, 1995.

a) NAFTA b) IMF

c) IRDB d) **D. GATT**

12. AFTA is ___________.

a) **ASEAN Free Trade Area** b) American Free Trade Area

c) Asian Free Trade Area d) Agreement for Free Trade Area

13. ASEAN stands for ___________.

a) The Association of Southeast American Nations

b) **The Association of Southeast Asian Nations**

c) The Agreement of Southeast American Nations

d) The Agreement of Southeast Asian Nations

14. __________ was established by a multilateral treaty of 23 countries in 1947.

a) WTO b) UN

c) **GATT** d) NAFTA

15. In International Trade, IMF stands for __________.

a) **International Monetary Fund**

b) International Money Fund

c) International Market Fund

d) International Monetary Firm

16. _______is a fixed percentage on the value of the traded commodity.

a) Anti dumping duty b) Specific tariff

c) **Ad Valorem tariff** d) A compound tariff

17. In most countries, foreign trade represents a significant share of ______

a) EXIM b) FDI

c) Income Per Capita d) **GDP**

18. ________is a combination of an ad valorem and specific tariff.

a) Anti dumping tariff b) Specified Valorem Tariff

c) EXIM Tariff d) **A compound tariff**

19. _______ refers to goods imported from one country and are exported to another country.

a) Third Party Trade b) **Entrepot trade**

c) Export Trade d) EXIM Trade

20. _____ provided a series of 'rounds' of negotiations by which tariffs were reduced.

a) IMF b) NAFTA

c) IBRD d) **GATT**

22. ______ happens when Imports are more than exports.

a) Trade barrier b) **Trade deficit**

c) Trade surplus d) Trade contract

23. ______ happens when Exports are more than imports.

a) Trade barrier b) Trade deficit

c) **Trade surplus** d) Trade contract

24. BOP stands for ________.

a) Bank of Payments b) Barrier of Payments

c) Bill of Payments d) **Balance of Payments**

25. The ________ is composed of capital account and current account.

a) Bill of credit b) Barrier of Payments

c) Bill of Payments d) **Balance of Payments**

26. The full form of UNCTAD is ______

a) United Nations Conference on Tariff and Duties

b) United Nations Committee on Trade and Development

c) **United Nations Conference on Trade and Development**

d) United Nations Council on Tariff and Development

27. _____ is the oldest trade theory.

a) Comparative Cost Advantage Theory

b) Hecksher Ohlin Theory

c) Product Life Cycle Theory

d) **Theory of mercantilism**

28. _____ suggests that each country should specialize in producing only those goods which it can produce efficiently.

a) Theory of mercantilism b) **Theory of absolute advantage**

c) Product Life Cycle Theory d) Hecksher Ohlin Theory

29. Theory of absolute advantage is propagated by ______.

a) Philip Kotler b) **Adam Smith**

c) Peter Drucker d) David Ricardo

30. Theory of comparative advantage was given by ______

a) Philip Kotler b) Adam Smith

c) Peter Drucker d) **David Ricardo**

31. __________ stresses on the 'relative cost difference' rather than 'absolute cost difference' between the goods as the basis for carrying out trade.

a) Theory of absolute advantage

b) **Theory of comparative advantage**

c) Hecksher Ohlin Theory

d) Theory of mercantilism

32. Factor endowments theory of international trade is also known as

a) **Theory of absolute advantage**

b) **Theory of comparative advantage**

c) **Hecksher Ohlin Theory**

d) **Theory of mercantilism**

33. _______ advocates that a country should focus and specialize in the production of goods that it can produce relatively at a lower cost than other countries.

a) Theory of absolute advantage

b) **Theory of comparative advantage**

c) Hecksher Ohlin Theory

d) Theory of mercantilism

34. International product life cycle theory was given by _________.

a) **Raymond Vernon** b) David Ricardo

c) Michael Porter d) Adam Smith

35. National competitive advantage also called as_____________.

a) Gold Model b) Silver Model

c) **Diamond Model** d) Platinum Model

36. According to 'New Trade Theory' a firm acquires export competitiveness due to _____________.

a) Specialization and economies of scale

b) Being the first mover in the market

c) Government support

d) **All of the above**

37. The theory of 'absolute advantage':

a) **There are two countries, A and B, and potentially two goods which can be traded, X and Y. A is absolutely better at producing X and B is absolutely better at producing Y and they trade together, then both countries will gain.**

b) The global strategy of businesses who always seek to gain an absolute advantage over their rivals.

c) Why developed countries have a competitive advantage over poorer countries.

d) A situation where there are two countries, A and B, and potentially two goods which can be traded, X and Y. If A was absolutely better at producing both X and Y compared to B then there would be no advantage in A trading with B.

38. According to the theory of comparative advantage, a country will export a good only if

a) It can produce it using less labor than other countries.

b) Its productivity is higher in producing the good than the productivity of other countries in producing it.

c) Its wage rate in producing the good is lower than in other countries.

d) Its cost of producing the good, relative to other goods, is at least as low as in other countries.

39. According to the theory of comparative advantage, countries gain from trade because

a) Trade makes firms behave more competitively, reducing their market power.

b) All firms can take advantage of cheap labor.

c) Output per worker in each firm increases.

d) World output can rise when each country specializes in what its does relatively best.

40. When did the World Trade Organization come into effect?

a) March 6, 1996
b) April 8, 1994
c) January 1, 1995
d) February 5, 1994

41. How many members are present in the WTO?

a) 207
b) 195
c) 160
d) 164

42. Where is the headquarters of the WTO located?

a) Austria
b) Geneva
c) New York
d) Washington DC

43. Which of these institutions is not a part of the World Bank Community?

a) IFC
b) IDA
c) WTO
d) IBRD

44. Along with the World Bank and ________________, WTO is the third economic pillar of worldwide dimensions.

a) International Economic Association (IEA)

b) International Monetary Funds (IMF)

c) International Development Bank (IDB)

d) International Funding Organisation (IFO)

45. Among the following options, which one is not the objective of the WTO?

a) To protect environment

b) To improve the balance of payment situation in the member countries

c) To improve the standard of living of people of the member countries

d) To enlarge production and trade of goods

46. Who is the current Director-General of WTO?

a) Pascal Lamy b) Mahmoud Riad

c) Chedli Klibi **d) Ngozi Okonjo-Iweala**

47. As a part of WTO guidelines, Agreement on Agriculture (AOAdoesn't consider)

a) Direct payments to farmers are permitted.

b) **Indirect assistance and support to farmers including R & D support by govt. are not permitted.**

c) Domestic policies which directly effect on production and trade have to be cut back.

d) Least developed countries do not need to make any cuts.

48. Which of the following is not the pillar of AoA

a) Enhancement of market access

b) **Protection of properties**

c) Reduction of domestic support

d) Elimination of export subsidies

49. IBRD (International Bank for Reconstruction and Development) also known as

a) EXIM Bank b) **World Bank**

c) IMF d) International Financial Bank

50. Trademark Act, came into force on

a) 1957 b) 2000

c) 1970 d) **1999**

51. Which of the following is not type of patent

a) Utility patents b) **Copyright**

c) Design patents d) Plant patents

52. Copyright Act, came into force on

a) **1957** b) 1999

c) 2000 d) 1970

53. Which of the following is Rights of a Copyright Owner

a) Publish their work and Perform their work in public

b) Translate and Broadcast their work

c) Prevent others from making unauthorized use of copyrighted work

d) **All of the above**

54. ____________ protects the intellectual property created by artists?

a) **Copyright** b) Geographical indications

c) Patents d) Registered designs

55. ______________ protects the intellectual property created by designers?

a) Copyright b) Trademarks

c) **Registered designs** d) Patents

56. ________________ protects the intellectual property created by inventors?

a) Copyright b) **Patents**

c) Registered designs d) Trademarks

57. What does a trademark protect?

a) An invention b) A work of art

c) **Logos, brands and names** d) Look, shape and feel of a product

58. In most countries, how long does copyright last for?

a) 50 years after the making of the work

b) 10 years after the making of the work

c) **50 years after the death of the person who created that work**

d) 10 years after the death of the person who created that work

59. How long the patents usually last for?

a) 10 years b) **20 years**

c) 40 years d) 60 years

60. If you write an original story, what type of intellectual property protection gives you the right to take decisions on who can make it again and sell the copies of your work?

a) **Copyright** b) Patents

c) Registered designs d) Trademarks

61. Imagine a sports team sets up a company to sell its own range of clothes. Which type of intellectual property protection can the team use to show that the clothes are made by them?

a) Copyright b) Geographical indications

c) Registered designs d) **Trademarks**

14

Information Technology in Marketing

1. **Market information includes**

   a) Market intelligence b) Market news

   c) Both a and b d) Marketing

2. **Historical nature of market information is**

   a) Market intelligence b) Market news

   c) Market information d) Marketing

3. **Current information about markets is**

   a) Market intelligence **b) Market news**

   c) Market information d) Marketing

4. **___________ defined as gathering, recording and analyzing of all facts relating to the purchase and sale of goods and services from producer to consumer**

   a) Market intelligence b) Market news

   c) Market information **d) Market research**

5. **It is not the step of market research**

   a) Hypothesis formulation

   b) Problem identification

   c) Data analysis

   d) Undesigning the empirical procedures

6. **______________ is broadly defined as a communication or, reception of knowledge or intelligence.**

   a) Market information b) Market news

   c) Market arrivals d) None of the above

7. ________________ describes research on markets, market size, geographical distribution and incomes

a) Market research b) Marker conduct

c) Market information d) Market news

8. Major objectives of market research

a) Understanding the existing traditional marketing system

b) Diagnosing the problems in a dynamic content

c) Analyzing and predict the impact and effectiveness of alternate policy measures

d) All the above

9. Which of the following is/are the components of market structure?

a) Concentration of market power

b) Degree of integration

c) Degree of product differentiation

d) All the above

10. The market information pertaining to market arrivals, prices, demand for commodities etc., in the past period refers to

a) Market news **b) Market intelligence**

c) Communication d) Market extension

11. The market information pertaining to market arrivals, prices, demand for commodities etc., in the current period refers to

a) Market news b) Market intelligence

c) Communication d) Market extension

12. ___________can be measured by a number of factors, such as the number of competitors in an industry, the heterogeneity of product and the cost of entry and exit

a) Market conduct b) Market intelligence

c) Communication d) Market news

13. ____________ is used to test the viability of a new product or service by communicating directly with a potential customer

a) Market information b) Marker intelligence

c) Market research d) Market news

14. The concentration of market power is an essential element in determining the nature of competition, _______________ and _______________

a) Market information and market news

b) Marker intelligence and market information

c) Market conduct and performance

d) Market news and market intelligence

15. _____________ is not the components of the market structure which together do not determine the conduct and performance of the markets

a) Concentration of market power

b) Degree of product differentiation

c) Conditions for entry of firms in the market

d) Technology

16. Marketing research is concerned with __________

a) Anticipation of production

b) Supply position

c) Financial problems

d) Solution to specific problems of marketing

17. Which one of the following companies has started a rural marketing network called e-chaupals?

a) Proctor and Gamble

b) Hindustan Lever

c) Dabur

d) ITC

18. _________________ is the electronic trading portal which links the APMC mandis to generate a unified national wide market for agricultural commodities in India

a) BARX

b) eAgriBazar

c) eNAM

d) REMS

19. ______________ is the leading agency for establishing eNAM with the sponsorship of Ministry of Agriculture and Farmers' Welfare, Government of India.

a) Small Farmers Agribusiness Consortium

b) Rural Agribusiness Consortium

c) Agribusiness Consortium for Rural Farmers

d) Indian Farmers Agribusiness Consortium

20. e-NAM is not aimed in

a) One Nation One Market

b) Improve price discovery mechanism

c) Eradicating information asymmetry

d) One Nation Multi Market

21. _______________ is an initiative by Indian Tobacco Company limited (ITC) in the year 2000

a) DEMIC **b) e-Choupal**

c) e-NAM d) IKisan

22. Nationwide information network project of Directorate of Marketing and Inspection (DMI) is

a) AGMARKNET b) IKISAN

c) e-NAM d) e-Choupal

23. National level network of information for rapid collection and diffusion of price, commodities, sales, arrival information, contract farming

a) e-Choupal b) AMI

c) AGMARKNET d) e-NAM

24. Find the odd one regarding to reforms in the APMC Act which are required to carry out by the states to integrate their mandis with e-NAM

a) Single trading license to be valid across the state

b) Single point levy of market fee across the state

c) **Multiple point levy of market fee across the state**

d) Provision for e-auction/ e-trading as a mode of price discovery

25. eNAM was established in the year

a) 2016 b) 2014

c) 2012 d) 2008

26. e-Choupal do

i) Leverages information technology

ii) Makes use of the physical transmission capabilities of current intermediaries

iii) Bridge financing

a) i, ii and iii are correct b) i and ii are only correct

c) i and iii are only correct d) ii and iii are only correct

27. DEMIC denotes

a) Directorate of Export and Import Market Intelligence Cell

b) Domestic and Export Market Intelligence Cell

c) Directorate of Export Market Intelligence Cell

d) Domestic Import and Export Market Intelligence Cell

28. Domestic and Export Market Intelligence Cell was established in

a) 2006 b) 2009

c) 2004 d) 2008

29. Which of the following was the first private sector initiative by the private sector in agricultural marketing?

a) E-Chaupal b) Trifed

c) Nafed d) None of the above

30. Information related to market facts about the prices that prevailed in the past period and also market arrivals over time is __________

a) Market News **b) Market information**

c) Market trend d) Market magazine

31. Up-to-date information about prices, arrivals and changes in market conditions is ______________

a) Market information b) Market trend

c) Market magazine **d) Market News**

32. External sources of marketing information is

a) Magazines

b) Commercial press

c) Annual Reports of the Trade Association

d) Trade journals

33. External sources data have been usually checked for reliability before it is entered into an marketing information system

a) Market intelligence **b) Marketing information**

c) Marketing survey d) Marketing research

34. A set of procedures and methods for the regu-lar, planned collection, analysis, and presentation of information for use in making marketing decisions

a) Marketing research

b) Market intelligence

c) Marketing survey

d) Marketing information system

35. Data on sales, inventories, marketing costs, cash flows, accounts receivables and payables, trading returns, financial returns

a) Internal Marketing Information

b) Marketing research

c) External Marketing Information

d) Market intelligence

36. Marketing research is defined as the collection and analysis of data relevant to marketing decision-making and the communication of the results of this analysis to marketers

a) Marketing research b) Market intelligence

c) Marketing survey d) Marketing information system

37. Facts, opinions, estimates, and all other information which affect the marketing of goods and services is

a) Market information b) Market Data

c) Market system d) e-market

38. Criteria for Good Market Information

a) Comprehensive b) Relevance

c) Disclosed d) Timeliness

39. External sources of marketing information is

a) Magazines

b) Commercial press

c) Annual Reports of the Trade Association

d) Trade journals

42. External sources data have been usually checked for reliability before it is entered into an marketing information system

a) Market intelligence
b) Marketing information
c) Marketing survey
d) Marketing research

43. A set of procedures and methods for the regu-lar, planned collection, analysis, and presentation of information for use in making marketing decisions

a) Marketing research
b) Market intelligence
c) Marketing survey
d) Marketing information system

44. Data on sales, inventories, marketing costs, cash flows, accounts receivables and payables (credit sales and credit purchases), trading returns, financial returns

a) Internal Marketing Information
b) Marketing research
c) External Marketing Information
d) Market intelligence

45. Facts, estimates, opinions, guidelines, policies and other data which is necessary for taking marketing decisions

a) Marketing system
b) Marketing information
c) Marketing survey
d) Marketing research

47. ____________ is defined as the collection and analysis of data relevant to marketing decision-making and the communication of the results from the analysis to marketers

a) Marketing information system
b) Marketing research
c) Market intelligence
d) Marketing survey

48. Which one of the following companies has started a rural marketing network called e-chaupals?

a) Proctor and Gamble
b) Hindustan Lever
c) Dabur
d) ITC

49. e-NAM is not aimed in

a) One Nation One Market

b) improve price discovery mechanism

c) Eradicating information asymmetry

d) One Nation Multi Market

50. It is an initiative by Indian Tobacco Company limited (ITC) in the year 2000

a) DEMIC **b) e-Choupal**

c) e-NAM d) IKisan

53. Find the odd one regarding to reforms in the APMC Act which are required to carry out by the states to integrate their mandis with e-NAM

a) Single trading license to be valid across the state

b) Single point levy of market fee across the state

c) Multiple point levy of market fee across the state

d) Provision for e-auction/ e-trading as a mode of price discovery

54. eNAM was established in the year

a) 2016 b) 2014

c) 2012 d) 2008

55. APMC was created to

a) Protect the interest of private sector in agriculture

b) Protect the interest of the farmers

c) Provide better suggestive measures to enhanced utilisation of green fertiliser

d) Protect the interest of the buyers of agricultural produce